6/98

ML

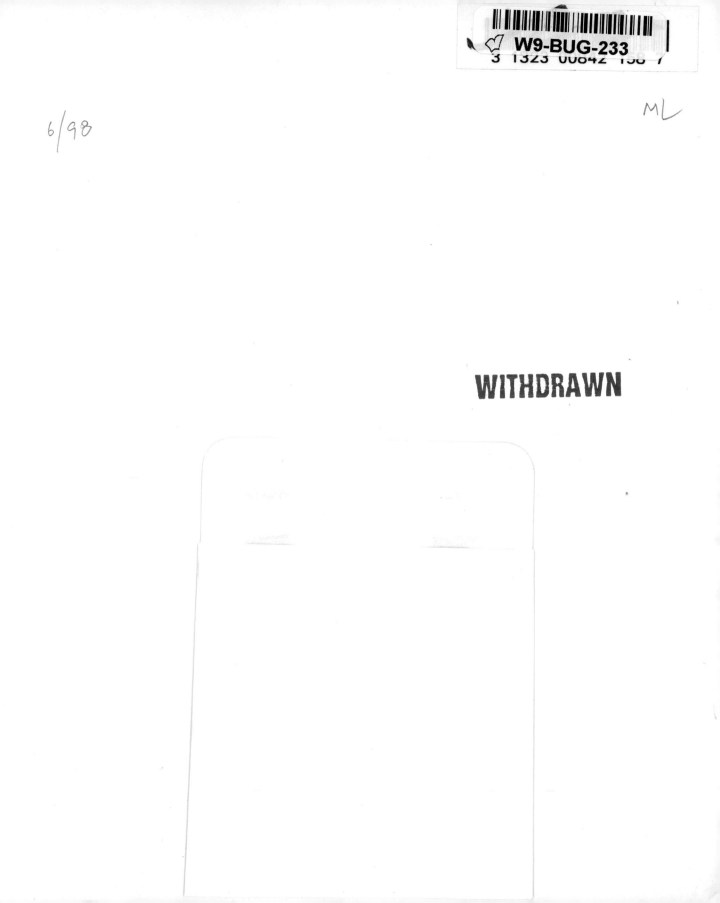

THE BRITISH LIBRARY
writers' lives

Charlotte Brontë

Vol. 3ʳᵈ

Chap. 23ʳᵈ

Bicknell

The Valley of the Shadow of Death

The Future sometimes seems to sob a low warning of the event
it is bringing us, like some gathering though yet remote storm which, in
tones of the mind, in flushings of the firmament, in clouds strangely
torn announces a blast ~~that will~~ strong to strew the sea with wrecks, or
~~that will~~ commissioned to bring in fog the yellow taint of pestilence, covering whole
western isles with the poisoned exhalations of the East, dimming the
lattices of English homes with the breath of Indian ~~Sunday~~ Plague.
At other times this Future bursts suddenly, as if a rock had rent,

THE BRITISH LIBRARY
writers' lives

Charlotte Brontë

JANE SELLARS

OXFORD UNIVERSITY PRESS

Contents

⬳ *Preface*

Charlotte Brontë's life story is almost as well-known as the history of her greatest heroine, Jane Eyre. Indeed, for many readers, the two become inextricably entwined. The Victorian audience was convinced that Charlotte's dedication of the second edition of *Jane Eyre* to her literary hero, William Makepeace Thackeray, meant that the novel was in fact the true account of Thackeray's own governess.

When Charlotte Brontë and her sisters, Emily and Anne, adopted the male pseudonyms of Currer, Ellis and Acton Bell for the publication of their *Poems* in 1846 and the classic novels which followed, they were already aware of the need to conceal their female identities. Later in life, her beloved sisters both gone, Charlotte took action to protect, as she saw it, Emily's and Anne's posthumous privacy and their literary reputations. At the same time she struggled with the public demands of her own fame, suffering rather than enjoying the glamorous rewards, often preferring to retreat into the dreary but comfortably familiar home life of Haworth Parsonage.

After Charlotte's death in 1855, her widower Arthur Bell Nicholls and her aged father, the Reverend Patrick Brontë, grieving for a much-loved wife and daughter, also had to come to terms with a world which clamoured to know more of the life of the famous author, Currer Bell. So, too, did Ellen Nussey, Charlotte's lifelong friend and recipient of decades of her letters. These three people together held the key to Charlotte's history, and each in their own way contributed to what has become an undying fascination with this author of genius. Each of them played a part in informing the first and best-known biography, Elizabeth Gaskell's *The Life of Charlotte Brontë*, published in 1857. Mrs Gaskell's *Life* ignited an interest in the Brontës which has today become a passion on an international scale.

This biography, *The British Library Writers' Lives: Charlotte Brontë*, draws many of its sources and illustrations from the outstanding collections of Brontë manuscripts and memorabilia in The British Library in London and the Brontë Parsonage Museum at Haworth in Yorkshire, formerly the Brontës' home and now a literary museum dedicated to their literature and their lives. It does not seek to pose new theories or make startling revelations about Charlotte Brontë and her art as a writer, but is intended as a fluent account of her life, an introduction to the extraordinary background which produced this much-loved genius of English literature.

This book is also a celebration. Published in 1997, it coincides with the 150th anniversary of one of the most significant events in the history of nineteenth-century literature: the publication of Charlotte Brontë's *Jane Eyre*, Emily Brontë's *Wuthering Heights* and Anne Brontë's *Agnes Grey*.

Acknowledgements

In writing this book I have been dependent both on my own accumulated knowledge of seven years as director of the Brontë Parsonage Museum at Haworth, and the work of generations of other writers on the Brontës. Thanks are also due to the Brontë Society, which generously provided many of the illustrations.

I dedicate this book to the memory of my father Colin Sellars (1925-1997), whose scholarly love of English literature has been my constant inspiration.

Jane Sellars
May 1997

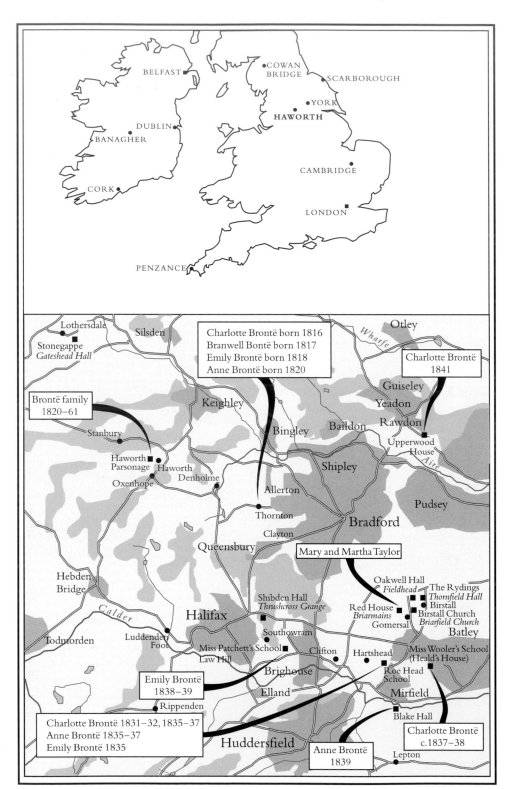

Some places in Great Britain and Ireland associated with the Brontë family.

BELFAST

COWAN BRIDGE

SCARBOROUGH

DUBLIN

BANAGHER

YORK

HAWORTH

CORK

CAMBRIDGE

LONDON

PENZANCE

Lothersdale

Silsden

Stonegappe
Gateshead Hall

Otley

Wharfe

Charlotte Brontë born 1816
Branwell Bontë born 1817
Emily Brontë born 1818
Anne Brontë born 1820

Charlotte Brontë
1841

Guiseley

Brontë family
1820–61

Keighley

Yeadon

Rawdon

Stanbury

Bingley

Baildon

Haworth
Parsonage

Haworth

Denholme

Shipley

Upperwood
House

Aire

Oxenhope

Allerton

Pudsey

Thornton

Bradford

Clayton

Queensbury

Mary and Martha Taylor

Hebden
Bridge

Calder

Shibden Hall
Thrushcross Grange

Oakwell Hall
Fieldhead

The Rydings
Thornfield Hall
Birstall

Red House
Briarmains
Gomersal

Birstall Church
Briarfield Church

Halifax

Southowram

Batley

Todmorden

Luddenden
Foot

Miss Patchett's School
Law Hill

Clifton

Hartshead

Miss Wooler's School
(Heald's House)

Emily Brontë
1838–39

Brighouse

Roe Head
School

Elland

Mirfield

Rippenden

Blake Hall

Charlotte Brontë 1831–32, 1835–37
Anne Brontë 1835–37
Emily Brontë 1835

Huddersfield

Anne Brontë
1839

Lepton

Charlotte Brontë
c.1837–38

⟋ *A Haworth Childhood*

Haworth is a bleak-looking place; a string of dark stone buildings, clinging and massing together on a precipitous Pennine hillside. At the top of the steep Main Street, cobbled with wide stone setts to stop the horses' hooves from slipping, a narrow lane turns a sharp corner and climbs up past the church, between the graveyard – overflowing with tombstones – and the low-roofed Sunday School building. At the top of the lane stands Haworth Parsonage, a neat, symmetrical building, with its nine many-paned windows overlooking a square of garden which abuts into the sombre churchyard, so that the house seems to rise from a sea of gravestones: ever-present reminders of death surrounding the place which became the Brontë family's home in 1820. Behind the parsonage the land sweeps away into the dramatic wildness of the moors, the harsh landscape which for the Brontës was their childhood playground and their literary inspiration.

In February 1820, the Reverend Patrick Brontë was appointed Rector of Haworth, and two months later he arrived to take up residence at the parsonage with his Cornish-born wife Maria and his six children: Maria, Elizabeth, Charlotte,

The earliest known image of Haworth Parsonage, an ambrotype which dates from the 1850s. The trees in the graveyard which now surround the parsonage were not planted until later that decade.

The Brontë Society

Branwell, Emily and Anne. Charlotte was just four years old and Anne was a babe-in-arms. The Brontës had come to Haworth from the parish of Thornton, Bradford, where all the children except Maria had been born, and where they had spent five pleasant, sociable years.

Haworth was Patrick Brontë's fourth position in the church since he graduated from St John's College, Cambridge in 1806. He made an extraordinary father – lively, intellectual, with interests in science, medicine, art, literature and music that reached far beyond his province as a rector of a small Yorkshire parish. Patrick Brontë was Irish, born in Emdale, County Down, in 1777, and he had clawed his way from his humble origins to acquire a hard-won education and a sponsored place at Cambridge University. He came to Yorkshire in the first place as minister at Hartshead Church near Dewsbury, and it was at nearby Woodhouse Grove School that he met Maria Branwell, who was visiting relatives there and came from faraway Penzance. After a brief courtship, Patrick and Maria married in December 1812. They came to Haworth as a young and lively family, optimistic for the future, prepared to set down roots in this inhospitable corner of the West Yorkshire landscape.

The Haworth the Brontës knew was a harsh place in more ways than one. Situated eight hundred feet high in the Pennines, the small township suffered an extreme winter climate of impenetrable dampness, fierce cold winds and blanketings

Haworth Parsonage and the church of St Michael and All Angels, Haworth, from a drawing by Elizabeth Gaskell.

The Brontë Society

of snow. Charlotte's hundreds of letters to her life-long friend Ellen Nussey contain innumerable references to the state of the weather and the effect on the family's health, health being the other preoccupying issue of life in Haworth. Living conditions for the crowded local population of this industrial centre were poor. There were no sewers and the water supply was inadequate and polluted, thus creating a high mortality rate. There were a scarcely believable 1344 burials in Haworth churchyard between 1840 and 1850, and the average age of death was a mere twenty-five years. Less than sixty per cent of babies born there survived beyond their sixth birthday. Although the Brontë deaths came tragically early, they seem unremarkable when set against this background.

Subsistence farming, hand-loom weaving and wool combing made up the local employment. When the Brontës arrived in Haworth the domestic system of worsted manufacture was changing to factory production with water-powered machinery and there were already long-established mills working alongside the river Worth. Quarrying, building and crafts were the only other areas of work at the time, though there were also a few professional people in the neighbourhood. The church and the flourishing Baptist and Wesleyan chapels provided the only education on offer, and they formed the focus of the community's social life.

Much has been made of the Brontës' isolation, of how they lived on the edge of the moors, outsiders in a coarse environment where life was nasty, brutish and short. Elizabeth Gaskell is largely responsible for this persistent view, because, in her *Life of Charlotte Brontë*, in an attempt to defend Charlotte from accusations of coarseness in the novels, she accentuated the primitive character of Haworth, basing her descriptions of the place on how it was in the eighteenth century. Her aim was to show that Charlotte's writing came out of innocence trapped in a savage environment. In fact, mid-nineteenth-century Haworth was a fast-developing town, standing at the apex of a group of substantial industrial centres – Halifax, Burnley and Keighley – and consequently the focus of much traffic of trade and people. There was too a self-sufficiency about Haworth – every process of the worsted trade was carried out there – which imbued the tough character of the local people. If the Brontës were isolated, it was socially rather than physically. As Patrick Brontë wrote in a letter to a friend in November 1821 of how he struggled to cope with his wife's

A portrait of Patrick Brontë as a young man by an unknown artist.

The Brontë Society

fatal illness without his friends around him, he felt like 'a stranger in a strange land'. Maria Brontë died on 15 September 1821, her six little children at her bedside. After his wife's death, Patrick Brontë, forty-seven years old, without fortune or social position and six young children to raise, found himself to be an unmarriageable prospect. It was the generosity of his deceased wife's sister, Elizabeth Branwell, which saved the Brontë household. She gave up her comfortable life in Cornwall, and any prospect of marriage for herself, to come and live at Haworth Parsonage to take on the role of housekeeper and surrogate mother.

Lurid tales abound of Patrick's severity as a father, again promoted by Mrs Gaskell, unwisely directed by the vengeful tales of a nurse sacked from the Brontës' employ. However, more reliable sources paint a more appropriate picture of the Brontës' childhood; a picture in accord with the early fostering of lively minds and incipient genius. One well-known story in particular, told by Patrick himself in a letter to Gaskell, demonstrates not only the children's precocity, but also Patrick's enthusiastic involvement in their personal and intellectual development. Charlotte Brontë was about eight years old at the time:

> '*thinking that they knew more, than I had yet discover'd, in order to make them speak with less timidity, I deem'd that if they were put under a sort of cover, I might gain my end – and happening to have a mask in the house, I told them all to stand, and speak boldly from under cover of the mask ... I then asked Charlotte, what was the best Book in the world, she answered, the Bible – and what was the next best, she answer'd the Book of Nature – I then asked what was the best mode of education for a woman, she answered, that which could make her rule the house well...*'

Domestic details of Charlotte Brontë's early life can be gleaned from the reminiscences of Sarah Garrs, who with her sister Nancy was a family servant until 1825. After morning prayers and a breakfast of porridge, milk, bread and butter, the children had lessons with their father in his study, followed by sewing instruction with Sarah until dinnertime at two o'clock, when they usually ate meat followed by milk-pudding. In the afternoon they walked out onto the moors, coming home for tea in the kitchen, followed by more lessons and discussion with their father, then night-time prayers and bed.

A sampler worked by Maria Brontë, the only surviving relic of Charlotte's eldest sister who died of consumption in 1825.

The Brontë Society

There was nothing unusual about the self-sufficiency of a large family of children close together in age. It was natural that they should need no other playmates than each other. All of them were voracious readers from an early age, and endless source material for their play was found in the local newspapers, the Leeds Mercury and the Leeds Intelligencer, with their accounts of political debate, tales of Charlotte's childhood hero the Duke of Wellington, reviews of books and magazines, high society gossip and local scandal and trivia.

When his children were still very young, Patrick was concerned for his daughters' futures. What were they going to do with their lives? Being realistic, he knew that the daughters of a poor clergyman had very few options open to them. Their marriage prospects were poor, careers in the professions were out of the question for women at the time, and working-class occupations could not be considered. All that was left was teaching. Whatever the girls did, they needed a reasonable education, whether it was to provide them with the feminine accomplishments needed to make them more attractive marriage prospects, or the qualifications to become governesses. Accordingly, Maria and Elizabeth, when nine and eight years old, were packed off to Crofton Hall boarding school near Wakefield. Their stay was brief, presumably because Patrick could not easily afford the fees. So he must have been delighted when, in December 1823, he came across a newspaper advertisement for a school for Clergymen's daughters at Cowan Bridge near Kirby Lonsdale, forty-five miles from Haworth. It was not just the fact that the school was cheap – the fees were only fourteen pounds a year, half as much as Crofton Hall – but the litany of reassuringly respectable names on the list of patrons who supported the Reverend Carus Wilson's school which convinced Patrick that this was the destination for his daughters.

The Clergy Daughters' School at Cowan Bridge, from an engraving of 1824, immortalised by Charlotte Brontë as the infamous Lowood School in Jane Eyre.

The Brontë Society

The description of Jane Eyre's traumatic period as a pupil at Lowood School which forms the opening chapters of the novel contain some of the most searing and affecting passages of Charlotte Brontë's writing. The reader cannot help but shudder at Brocklehurst's humiliation of the orphan Jane Eyre, at the sickening accounts of poor and inadequate food, at the ordeals of Sunday:

> *'Sundays were dreary days in that wintry season. We had to walk two miles to Brocklebridge Church, where our patron officiated. We set out cold, we arrived at church colder: during the morning service we became almost paralysed. It was too far to return to dinner, and an allowance of cold meat and bread, in the same penurious proportions observed in our ordinary meals, was served round between the services.'*

Clearly Charlotte was writing from the heart, and there can be no doubt that the fiction is based on the Brontë sisters' terrible experience of Cowan Bridge School. Charlotte suffered great hardship there: she hated being torn away from a loving home to suffer the communal school life, she fought against her loss of freedom, she felt ignobled by being labelled a charity child, and most of all she could never forget or forgive the fact that her two older sisters died as a direct consequence of their time at the school. It was the death of her beloved eldest sister, Maria, that inspired the creation of the almost saintly character of Helen Burns. Ironically, the fictional character was described by some critics as unbelievable. Charlotte Brontë was later swift to correct this:

> *'she was real enough: I have exaggerated nothing there; I abstained from recording much that I remember respecting her, lest the narrative should sound incredible. Knowing this, I could not but smile at the quiet, self-complacent dogmatism with which one of the journals lays it down that "such creatures as Helen Burns are very beautiful but very untrue".'*

The Children's Friend *for 1826, one of the books of cautionary tales for children written by Dr Carus Wilson, the founder of Cowan Bridge School.*

The Brontë Society

In February 1825, Maria became very ill with consumption and Patrick Brontë came to fetch her home at once. Charlotte, Elizabeth, and six-year-old Emily were left behind at the school. They were never to see their sister again, as Maria died at home on 6 May 1825, only eleven years old. Soon afterwards, Elizabeth Brontë succumbed to the same fatal disease which had taken her sister. She, too, came home to die at Haworth Parsonage, on 15 June 1825, aged ten years. Patrick Brontë then wasted no time in bringing Charlotte and Emily back to the family home, a homecoming made traumatic by the loss of their two beloved sisters.

Charlotte was then cast into the role of eldest Brontë. Her sense of loss stayed with her all of her life, and invades her writing in the form of the constant image of the orphan child. The deaths of her sisters were all the more bitterly felt when seen in the perspective of the religious teaching at Cowan Bridge, where the emphasis was on sin and the inevitability of punishment by death. There Charlotte had been fed the sinister, warning children's tales written by Carus Wilson. A typical story begins:

'Do look at that bad child. She is in a pet. She would have her own way. Oh! how cross she looks. And oh! what a sad tale I have to tell you of her. She was in such a rage, that all at once God struck her dead. She fell down on the floor, and died. No time to pray. No time to call on God...'.

When Charlotte wrote of Lowood School, some part of her was taking revenge on this heartless and insensitive doling out of Calvinist doctrine to vulnerable children.

All girls were entered into the register at Cowan Bridge with a comment on their capabilities. Charlotte's assessment describes a seven-year-old who:

Charlotte Brontë's earliest extant manuscript, one of the famous little books, measuring just 28 x 36 mm. It contains six tiny water-colours illustrating a story about Anne Brontë.

The Brontë Society

'Reads tolerably – Writes indifferently – Ciphers a little and works neatly. Knows nothing of Grammar, Geography, History or Accomplishments. Altogether clever of her age but knows nothing systematically.'

Her education, and that of her brother and sisters, after the rigours of Cowan Bridge, was from 1825 until 1830 carried out at home in a manner that continued to be unconventional. But it was possessed of a richness that provided Charlotte Brontë with the opportunities for a corporate juvenile creativity which formed the foundations of her later literary achievements.

During these important early years Charlotte was reading voraciously, writing stories, performing plays and learning to draw. No more than one would expect of a bright and creative child, but Charlotte and her siblings went beyond mere play by creating their own imaginary worlds, using inspiration from every source available to them to assist their creation. For example, they had the usual educational texts at their disposal, such as J. Goldsmith's *A Grammar of General Geography*, but their copy is heavily annotated, a clue to their less orthodox use of the book as a source for maps and place names for their invented lands. They had the favourite books of the period, John Bunyan's *Pilgrim's Progress*, John Milton's *Paradise Lost*; they had access to their father's classical texts from university, Homer, Horace, Lempriere's *Bibliotecha Classica*; and they had Aunt Elizabeth Branwell's *Lady's Magazines* and Methodist magazines. The Bible they knew inside out. Patrick Brontë borrowed books about science from the Keighley Mechanics' Institute, and the family no doubt subscribed to the circulating libraries of the time, providing them with history, biography, travel books, poetry and novels. But it seems to have been the monthly journal *Blackwood's Edinburgh Magazine* which had the most profound effect on Charlotte and her brother and sisters. It was from its pages that they learnt about contemporary politics and literature. Such precocious reading went hand-in-hand with normal childhood favourites, such as Aesop's *Fables* and tales of the Arabian Nights. Another staunch favourite was Thomas Bewick's illustrated *History of British Birds*, from whose pages Charlotte laboriously copied the studies of birds in all their detail, and the amusing, sometimes gruesome little wood-engraved vignettes that peppered the pages. In 1829, Patrick hired the Keighley artist John Bradley to give his children art lessons. Charlotte displayed a passion for drawing and laboured over her work to the extent that she conceived an early ambition to be an artist.

Blackwood's Young Men's Magazine, *the second issue of the tiny journal which Charlotte, with her brother and sisters, produced to chronicle the events in the world of the 'Young Men', the heroes inspired by Mr Brontë's gift of toy soldiers.*

The Brontë Society

Reading, drawing and play-acting, using their toy soldiers, the 'Young Men', as characters, led the young Brontës naturally into their writing and thus into the imaginary worlds of Glasstown and then Angria. Battles, revolution and politics dominated these worlds. Famously, these adventures were described in the Brontës' tiny books. Paper was scarce in the Brontë household, and the little books were made up of whatever the children could lay hands on, such as scraps of sugar bags and wrapping-paper. Just a few inches square, the little books necessitated the evolution of a minuscule handwriting, designed to look as though it had been printed. Because none but the children could read it, the tiny writing took on the significance of a secret code, so that adults were barred admittance to their imaginary worlds. The Brontës' little books are in any case very difficult for anyone else to read: Brontë scholars have had a perpetual struggle to decipher not only the miniature writing, but also the haphazard spelling and random punctuation.

The characters inspired by the toy soldiers, with their peculiar babyish names, such as Sneaky, Gravey and Waiting Boy, gradually evolved into robust adventurers based on contemporary heroes and villains such as Napoleon Bonaparte, the Duke of Wellington and the explorers Ross and Parry. Exotic locations for their adventures were found in reports published in *Blackwood's Edinburgh Magazine*. Magic, mystery and the supernatural crept into Charlotte's stories from the *Arabian Nights* and Celtic folk stories. She revelled in writing long passages of scene-setting description, creating vast marble palaces set with precious jewels and surrounded by olive groves, palm trees and the heady scent of Mediterranean flowers. Instigated by Branwell, the little books developed into more ambitious facsimile magazines, then newspapers. It is apparent from these early writings that a power struggle existed between Charlotte and her brother Branwell, as each jockeyed for the lead in artistic invention. Appropriately, their attentions gradually turned away from the soldier-hero to the artist-hero, and the writing proliferated with time: eighteen little books survive from 1829 alone.

In 1830, after five years of freedom at home, the spectre of Charlotte Brontë's future again raised its head. The eldest daughter, now fourteen years old, had had only one year of formal schooling in her life, and the day when she would need to earn a living was looming very close. Another chapter in her life was about to open .

Charlotte Brontë's copy of an engraving by Thomas Bewick, also the source for one of Jane Eyre's visionary pictures: 'a cormorant, dark and large, with wings flecked with foam; its beak held a gold bracelet set with gems, that I had touched with as brilliant tints as my palette could yield and a glittering distinctness as my pencil could impart',

Jane Eyre, *Chapter 13.*

~ *Education*

The importance of education in Charlotte Brontë's life cannot be over estimated. Her quest for knowledge is a key to her success as a writer. Patrick Brontë was an exceptional father in that he sought out an education for his daughters in an age which saw marriage as women's vocation, and which did not acknowledge the need for education for girls beyond domestic training and rudimentary accomplishments in reading, drawing and music. It was her access to education that fostered Charlotte Brontë's genius. The Reverend Brontë realised that his girls must equip themselves to be teachers, because, lacking fortune and good connections, the Brontë sisters were not easily going to find husbands to support them. So Mr Brontë took a pragmatic view of their situation. At the same time, he placed a high value on education for its own sake, as a means of broadening the mind and opening up choices. Patrick Brontë's respect for education sparked off in his highly intelligent children a thirst for art and literature, a craving which laid the foundations of the wide literary knowledge that invests all of Charlotte's writing with its dense allusions.

In January 1831, seven years after the tragic Cowan Bridge School episode, Charlotte went away from Haworth once more to study at Roe Head school, Mirfield, twenty miles from the parsonage. A drawing she made of Roe Head survives, depicting a view from the front gate. It is a large, double-fronted house with three floors, set in extensive grounds. The sweeping landscape of the increasingly industrialised Calder valley formed the backdrop to the school. Miss Margaret Wooler and her sisters Catherine, Susan, Marianne and Eliza ran the school, whose small population of just ten pupils consisted of the teenage daughters of wealthy local manufacturers. On her arrival at Roe Head, Charlotte met the two young women who were to be

A pencil drawing of Roe Head school made by Charlotte c.1831–1832. Anne Brontë made a study of exactly the same view of the building when she was a pupil there.

The Brontë Society

By my D. Daughter Charlotte Brontë Minr of Haworth

her life-long friends and correspondents – Ellen Nussey and Mary Taylor. These two left us descriptions of Charlotte's rather plain appearance, something of which she was herself acutely aware, particularly when exposed to her peer group for the first time. Mary Taylor described her as looking like a little old woman, physically frail and dressed in unflattering, old-fashioned clothes, peering through her spectacles like a little mouse, with her nose permanently stuck in a book. Although odd-looking, and doubly marked out for the strong Irish accent of her speech, Charlotte impressed her fellow pupils from the beginning with her outstanding knowledge of poetry and her love of pictures and drawing. She soon displayed a passion for learning, avidly embracing her lessons in geography, history, English grammar, French, music and drawing. The girls worked from a standard school text of the day, Richmal Mangall's *Historical and Miscellaneous Questions*. Learning by heart was the method, and Charlotte did well. After just six months she was easily top of the class, and won the medal for achievement, inscribed 'Emulation' on one side and 'Rewarded' on the other.

As she pored over her meticulously drawn copies of classical heads and learnt her French verbs, Charlotte missed the close companionship of her sisters and brother, and most of all the chance to exercise her literary imagination in weaving the shared tales of Glasstown, the fictional setting of their tales. But her intense application to her school-work indicates that she was determined to do her very best in order to improve her chances of finding a teaching situation and the means of earning a living to help her family. Art sustained her in her doggedly hard work, and every morsel of information about painting, architecture and music that came her way was squirreled away for future contemplation.

The recollections of her school-friends paint a picture of the teenage Charlotte Brontë as a very serious-minded girl who could not easily be swayed away from her studies into enjoying herself.

However, she was discovered to have a talent as a captivating story-teller, so much so that she quite terrified her friends with her lurid tales, and immediately suffered remorse for doing so.

The friendships Charlotte made at Roe Head were to be important to her for the rest of her life, and she is distinguished from her sisters in her ability to make and keep the friends of her school-days. She went home to Haworth only at Christmas and in the summer, and at weekends and other holidays she paid duty visits to friends of the family in the Dewsbury neighbourhood, which she endured, and visited the homes of her friends Ellen and Mary, which she enjoyed. Ellen's family, the Nusseys, were cloth-manufacturers who lived at the Rydings in Birstall. The Taylors lived at the Red House at Gomersal, a lively family home dominated by Joshua Taylor, manufacturer, banker and a Radical in his views, providing a stirring opposite to Charlotte's Tory background.

Charlotte made these studies of noses at Roe Head in 1831. Such exercises were typical of the way drawing was taught to young ladies, by copying from instructional manuals and prints.

The Brontë Society

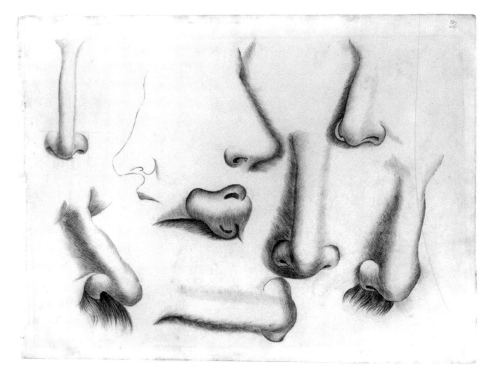

Ellen and Mary satisfied two different sides of Charlotte's personality. In later life Ellen was to be dependent on her brothers for her living of genteel poverty, whereas Mary emigrated to New Zealand and earned enough to live an independent retirement. Although we have a lifetime's letters from Charlotte to Ellen, we have no surviving letters to Mary. Ellen was the non-intellectual, conformist confidante, an emotional prop for Charlotte, someone in whom she could confide her worries and with whom she could comfortably discuss the domestic side of life. With Mary she could talk about books and politics and enjoy the cultured setting of her home, whilst at the same time Charlotte must have been conscious of the contrasting domestic poverty of her own Haworth household.

School life dominated Charlotte's world for only a brief period. Little imaginative writing survives from the eighteen months she spent at Roe Head, indicating that she was more preoccupied with the external than the internal world. But it was not to last, because by June 1832 she had progressed so fast in her year and a half of schooling that she left Roe Head to come home and teach her sisters.

The summer of 1832 saw the opening of the National Sunday School in Haworth, where Charlotte was also expected to teach. Her life took on a humdrum quality, something which she was to rail against for the rest of her days. She wrote to Ellen a description of her daily routine:

> ' *In the Morning from nine o'clock till half-past twelve – I instruct my Sisters*
> *& draw, then we walk till dinner after dinner I sew till tea time, and after*
> *tea I either read, write, do a little fancy-work or draw, as I please. Thus in one*
> *delightful, though somewhat monotonous course my life is passed.'*

Nearly fifteen years later, Charlotte Brontë put her feelings about enforced female domestic tranquillity into more succinct words in her novel *Jane Eyre*:

> ' *Women are supposed to be very calm generally: but women feel just as men feel;*
> *they need exercise for their faculties, and a field for their efforts as much as their*
> *brothers do; they suffer from too rigid a restraint, too absolute a stagnation,*
> *precisely as men would suffer; and it is narrow-minded in their more privileged*
> *fellow-creatures to say that they ought to confine themselves to making puddings*

and knitting stockings, to playing on the piano and embroidering bags. It is thoughtless to condemn them, or laugh at them, if they seek to do more or learn more than custom has pronounced necessary for their sex.'

Bolton Abbey, copied from a print after a drawing by JMW Turner. Charlotte exhibited this picture, along with Kirkstall Abbey, at the 1834 exhibition of the Royal Northern Society for the Encouragement of the Fine Arts in Leeds.

The Brontë Society

At this time, Charlotte's aura of discontent debarred her from entering wholeheartedly into the fictional world she had created with her brother Branwell. In 1832 she was concentrating on her poetry, writing carefully with revisions, as if for publication, for example *Lines on the Celebrated Bewick*, her early childhood artist-hero. Her heart was with her school friends, and she was happiest when visiting Ellen, or writing the earliest of the many hundreds of letters to her friend, which were later to tell her biographers so much about the patterns of her life, and sometimes little about her innermost feelings. For example, she opened the new year of 1833 with a letter of singular pomposity:

'The first day of January always presents to my mind a train of very solemn and important reflections and a question more easily asked than answered frequently occurs viz: How have I improved the past year and with what good intentions do I view the dawn of its successor? these my dearest Ellen are weighty considerations which (young as we are) neither you nor I can too deeply or too seriously ponder...'

However, not long after this sombre reflection, Charlotte found herself back in Glasstown again, competing with Branwell in literary creation. The influence of Walter Scott was strong as Charlotte wrote about the Marquis of Douro's first love affair, before his marriage to Marian Hume. All of her heroines were beautiful, aristocratic, proud, passionate and brave. She revelled in the embellishments of supernatural voices and the complexities of secret identities. *The Foundling*, written

Kirkstall Abbey, copied from a print by Charlotte in 1834.

The Brontë Society

THE FOUNDLING
A TALE OF OUR OWN TIMES BY
CAPTAIN
TREE

PRINTED AND PUB-
LISHED
BY
SEARGEANT TREE HOTEL
VERDOPOLIS

THIS BOOK

*The Foundling,
A Tale Of Our Own Times
by Captain Tree.
At the age of seventeen,
Charlotte was still deeply
involved in the imaginary
world of her childhood,
and continued to
collaborate with Branwell
and write in her
minuscule hand-writing.*

*The British Library
Ashley MS 159, f.1*

in May and June 1833, contains many of these elements, ones that were to be reworked many years later in *Jane Eyre*. A great deal of writing was going on in the parsonage at that time, all of it in the minuscule hand which the Brontës adopted in order to keep their world secret from adult eyes, and, more mundanely, to save on precious paper. Patrick Brontë, disturbed by the tortuous hand-writing, gave Charlotte a manuscript book inscribed 'All that is written in this book must be in a good, plain and legible hand.' A few poems found their way onto its pages, but Charlotte was too busy creating Angria, a new kingdom, to persevere. The stories became more and more complex and lurid. The Duke Zamorna had his child-bride killed off, married Mary Percy, then took up immoral behaviour, fathering a bastard child by a Negress, an evil dwarf called Finic. A first wife was revealed, with an illegitimate son.

In 1834 Charlotte reached her eighteenth birthday. She was still very much immersed in her writing, gradually fine-tuning her skills, and turning her attention nearer home for her inspiration, writing sharp caricatures of her brother Branwell in the character of Benjamin Wiggins who 'As a musician was greater than Bach; as a poet he surpassed Byron; as a painter Claud Lorrain yielded to him...', and dismal descriptions of Haworth 'a miserable little village buried in dreary moors and moss-hags and marshes.' When Ellen went off on a visit to London Charlotte burned with envy, and instead had to content herself with such village excitements as the installation of a new organ in Haworth church, concerts in Halifax and lectures at the Mechanics' Institute in Keighley. And there were the refinements her father managed to afford for their home, a cottage piano and prints of apocalyptic Biblical scenes by the artist John Martin. Indeed, at this time art was a preoccupying feature of Charlotte's life. Twenty-five drawings and paintings survive from 1834, many of them the highly finished productions of a young woman engaged in serious art study. The most significant of these is a pair of minutely detailed pencil copies of prints of picturesque views of two Yorkshire abbeys, Bolton Priory and Kirkstall Abbey. These pictures were selected to appear in the summer exhibition of the Royal Northern Society for the Encouragement of the Fine Arts in

Leeds that year. It has long been supposed that the Brontë family visited the exhibition with the prime objective of seeking out the Leeds artist William Robinson to tutor Branwell in his quest for a career as a professional painter, but they must too have gone to gaze with pride on Miss Brontë's drawings, hanging there on the wall alongside works by Richard Westall, William Mulready and Sir Thomas Lawrence.

Clearly, Charlotte Brontë did cherish an ambition to be a professional artist, but her hopes were against the odds of the discrimination against women artists at the time. Also, she must have realised, even then, that her own art education was so meagre that there was little chance of finding professional success. Where Branwell, the boy of the family, was given expensive lessons in oil painting, Charlotte never progressed beyond pencil, ink and water-colour. Years later, when asked in a letter of 1848 from her publisher to illustrate the second edition of *Jane Eyre* herself, Charlotte replied:

> '*It is not enough to have the artist's eye, one must also have the artist's hand to turn the first gift to practical account. I have, in my day, wasted a certain quantity of Bristol board and drawing-paper, crayons and cakes of colour, but when I examine the contents of my portfolio now, it seems as if during the years it has been lying closed some fairy has changed what I once thought sterling coin into dry leaves, and I feel much inclined to consign the whole collection of drawings to the fire.*'

Portrait making was a Brontë preoccupation around this time, with Anne Brontë a compliant model for Charlotte in both water-colour and pencil studies. But the best known image produced in 1834 came from the hand of the seventeen-year-old Branwell, his portrait of *The Brontë Sisters*, now one of the best-loved pictures in the National Portrait Gallery in London. It portrays the three young girls who were later to write some of the greatest works of English literature, each of them staring raptly out of the shadows. The canvas is scarred with the marks of fifty years spent folded up in an Irish wardrobe, where Arthur Bell Nicholls, Charlotte's widower, had stashed the canvas, considering it an unworthy representation. Nonetheless, the picture has a haunting quality which captures the intensity of expression of these

This idealised study of a woman's head is inscribed 'Amelia Walker', the name of one of Charlotte's pupils when she was a teacher at Roe Head. Amelia kept the drawing as a memento of her teacher.

The Brontë Society

Opposite page:

Branwell Brontë's famous portrait of his teenage sisters, Anne, Emily and Charlotte, painted in 1834 and now one of the most popular pictures in the National Portrait Gallery in London.

National Portrait Gallery, London

extraordinary women. The brown pillar in the centre of the group is where Branwell erased his own image early on in the painting process, not up to the technical skills required to make a successful composition of four figures.

In July 1835 Charlotte was compelled to start earning a living, and so it was back to Roe Head school, this time as a teacher, taking Emily with her as a pupil. Branwell, meanwhile, failed to make the grade even to apply to the Royal Academy of Arts in London as planned, and stayed at home turning out pedestrian portraits of Haworth tradesmen. Emily's stay at Roe Head was brief as she pined for home, so the stoical Anne took her place instead. Charlotte had no choice, and struggled to reconcile herself to her life as a school-teacher whilst all the time her head was full of the desire to write. She found the work increasingly tedious, and longed to be at home where she could indulge her literary fantasies in peace, without the interruption of the endless school routine. Whilst Branwell bubbled with a new plan to go off on the Grand Tour, never realised because of the usual lack of funds, his older sister yearned for her freedom. Her letters to Ellen sometimes betray her struggle with the demands of her job. In October 1836, after sixteen months of Roe Head, Charlotte described herself as

'Weary with a day's hard work – during which an unusual degree of Stupidity has been displayed by my promising pupils I am sitting down to write a few hurried lines to my dear Ellen. Excuse me if I say nothing but nonsense, for my mind is exhausted and dispirited. It is a Stormy evening and the wind is uttering a continual moaning sound that makes me feel very melancholy'.

As the year 1836 came to an end, both Charlotte and Branwell were making concerted efforts to progress their writing careers by writing for advice to admired literary figures. Branwell, careless as ever with his arrogant tone, even when addressing such luminaries, had no success in eliciting a reply from either William

Wordsworth, to whom he wrote in January 1837, or the editor of *Blackwood's Edinburgh Magazine*. Charlotte had more success. On 29 December 1836 she wrote a letter to the Poet Laureate, Robert Southey, seeking advice for an aspiring woman writer and enclosing some of her poems. The letter is lost, but Southey's reply gives some indication of Charlotte's approach. She must have blushed to have her girlish words quoted back at her, especially the phrase which describes Southey as 'stooping from a throne of light and glory'. He praises her gifts as a poetess, but cautions her in her ambitions – 'Many volumes of poems are now published every year without attracting public attention' – and implores her to write poetry for pleasure alone. The most quoted passage in the letter comprises perhaps the most famous piece of mis-advice in the history of English literature:

> *'Literature cannot be the business of a woman's life, & it ought not to be. The more she is engaged in her proper duties, the less leisure will she have for it, even as an accomplishment & a recreation. To those duties you have not yet been called, & when you are you will be less eager for celebrity.'*

Charlotte takes it very well, and is moved to reply to Southey to thank him for his kind words. She compares his advice to that of her own father 'who from childhood has counselled me just in the wise and friendly tone of your letter – I have endeavoured not only attentively to observe all the duties a woman ought to fulfil, but to feel deeply interested in them. I don't always succeed, for sometimes when I'm teaching or sewing I would rather be reading or writing...'. Certainly, she was not deterred in her writing, and produced sixty poems between January 1837 and July 1838, moving away from long narrative poems towards shorter, lyrical pieces.

The year 1838 was one of many comings and goings at the parsonage. Anne became ill and left Roe Head school, Branwell went off to Bradford to set himself up for a meagre year as a portrait painter, Emily had another attempt on the world of work with a short spell as a teacher at Miss Patchett's school at Law Hill, Halifax, and Charlotte did her best to resign herself to teaching at Miss Wooler's school, now removed to Healds House, Dewsbury. But by the end of the year she could take no more and left her post for good.

❧ *The Governess*

One of the reasons why *Jane Eyre* hit such a chord with the Victorian readership when it was first published in 1847 was that the plight of the governess was a cause for public concern in mid-nineteenth-century England. Although the 1841 census did not differentiate between school-teachers and governesses in private employment, the 1850 census recorded twenty-thousand governesses. There was a surfeit of single middle-class women in the second half of the nineteenth century, a social problem that was in itself a contradiction of the Victorian notion that marriage was women's only vocation. As the daughter of an impoverished clergyman, Charlotte Brontë fell into the problem category, a woman who was destined to become a governess. Indeed, even when Patrick Brontë

The Governess by Rebecca Solomon, from an engraving by Alfred T. Heath in The Keepsake, *1856. The image describes the isolation of the governess's situation which Charlotte found so unbearable.*

The Brontë Society

registered his girls, then all less than ten years old, at the infamous Cowan Bridge school, a note in the school register confirms that the purpose of their education was to be governesses.

By her very nature, the Victorian governess was a social incongruity. Elizabeth Rigby, later Lady Eastlake, writing in the *Quarterly Review* of December 1848 a joint review of *Jane Eyre* and *Vanity Fair*, emphasised the fact that the governess had no equals and no sympathy:

> '*She is a bore to most gentlemen, as a tabooed woman, to whom he is interdicted from granting the usual privileges of the sex, and yet who is perpetually crossing his path. She is a bore to most ladies by the same rule, and a reproach, too – for her dull, fagging, bread-and-water life is perpetually putting their pampered listlessness to shame. The servants invariably detest her, for she is a dependent like themselves, and yet, for all that, as much their superior in other respects as the family they both serve. Her pupils may love her, and she may take the deepest interest in them, but they cannot be her friends.*'

Charlotte Brontë never forgot the stigma of being a governess, and when as a celebrated author herself she visited the home of William Makepeace Thackeray, the author of *Vanity Fair* and her literary idol, she is reported to have spent most of the evening talking to the governess. In *Jane Eyre* she exorcised her feelings of resentment about the callous treatment of the lonely governess in the scene where a mortified Jane has to listen to Blanche Ingram mockingly recounting past displeasure with her governesses:

> ' *I have just one word to say of the whole tribe; they are a nuisance. Not that I ever suffered much from them; I took care to turn the tables. What tricks Theodore and I used to play on our Miss Wilsons, and Mrs Greys, and Madame Joubert! Mary was always too sleepy to join in a plot with spirit. The best fun was with Madame Joubert: Miss Wilson was a poor, sickly thing, lachrymose and low-spirited, not worth the trouble of vanquishing, in short;*

The Governess

A letter from
Charlotte to Ellen
Nussey, written on 5
and 6 December 1836.
Charlotte drew great
comfort from her
correspondence with
Ellen, during her
unhappy years as
teacher and governess.

The Brontë Society

and Mrs Grey was coarse and insensible; no blow took effect on her. But poor
Madame Joubert! I see her yet in her raging passions, when we had driven her
to extremities...'.

The governess's pay varied, but the usual rate was around thirty-five pounds
per year, although some were paid as little as sixteen or twelve pounds. When
Charlotte worked for the Whites of Upperwood House, Rawdon, she earned twenty
pounds, less four pounds for her laundry. Her sister Anne, although younger and less
experienced, was then earning twice as much with the Robinsons at Thorp Green
near York. For this low pay, many qualifications were expected, including the ability
to teach French, music, drawing, English, geography and needlework, at least. She
was banished to remote rooms at the top of house, forced either to eat alone in the
schoolroom, or, worse, with the children. A governess had to walk a little behind her
employers in public, she was not allowed visitors and her duties were hopelessly
undefined. If she did have spare time she was expected to do the household's sewing
all evening, as Charlotte discovered in her post with the Sidgwick family in 1839. She
may well have found herself looking after the baby, or reading aloud to the children's
mother. Over all this hung the constant fear of dismissal and the prospect of no
savings for later life. As Anna Jameson, contemporary writer and critic put it, all she
could look forward to was 'a broken constitution and a lonely unblessed old age'. It
was therefore not surprising that a high number of governesses ended up in lunatic
asylums.

In 1841 the Governesses' Benevolent Association was founded, and it
proceeded to change things in an intelligent manner. A fund was set up to help
women out of work, a hostel for them to live in, a bank to handle their financial
affairs, homes for the old, the ill and the retired, annuities for old age and a training
college, Queen's College, was founded in 1848. Eventually governessing became a
profession rather than a misfortune.

Charlotte Brontë did not have a vocation to be a teacher. Much as she valued
learning, it was for herself that she wanted the rewards, certainly not the girls in her
charge at Miss Wooler's school. Whilst working there, Charlotte wrote a telling

journal, using the minuscule hand-writing of her childhood, an indication that these were her private thoughts for herself alone:

> *'I had been toiling for nearly an hour with Miss Lister, Miss Marriott & Ellen Cook striving to teach them the distinction between an article and a substantive. The parsing lesson was completed, a dead silence had succeeded it into the school-room I sat sinking from irritation & weariness into a kind of lethargy. The thought came over me am I to spend all the best part of my life in this wretched bondage, forcibly suppressing my rage at the idleness the apathy and the hyperbolical & most asinine stupidity of these fat-headed oafs and on compulsion assuming an air of kindness, patience and assiduity ?'*

Several years later she was to discover that there was an even more wretched bondage, when in May 1839 she went off to her first post as governess with the Sidgwicks at Stonegappe, Lothersdale. Her innate pride and the sense of her own intellectual superiority to that of her paymasters added to the burden of teaching, the loathing for the feelings of subordination to her employers. The first months of the year Charlotte had spent at home, enjoying the companionship of her family, and turning down the unwanted attention of Ellen Nussey's cleric brother, Henry, who proposed marriage to her, much in the style of her later fictional creation, St John Rivers. Anne left in March to become governess with the Ingham family at Blake Hall, Mirfield, and Branwell was in Bradford in his portrait studio, enjoying the bohemian side of city-life. Charlotte's post at Stonegappe was a temporary one, but even then she found it hard to endure. She was affronted by the realisation that not only did she have to teach

A page from Charlotte Brontë's Roe Head journal, written in her minuscule hand-writing, in which she expressed her unhappiness about the lot of a school-teacher.

The Brontë Society.

the 'riotous, perverse, unmanageable cubs' but also to wipe their dirty noses and tie their shoe-laces. Her humiliation was almost complete.

It was with the Sidgwicks that Charlotte went to stay at Swarcliffe near Ripon, not far from Norton Conyers, a Jacobean house then rented by friends of her employers' family. It was then that Charlotte may have heard the legend of the madwoman in the attic at Norton Conyers, a tale she used in *Jane Eyre*. From Swarcliffe she wrote a miserable catalogue of a 'Private Governess's trials and crosses in her first Situation' in a letter to Ellen of 30 June. By the following month she had left the Sidgwicks' employment.

The years 1839 and 1840 were beset with the attentions of curates for Charlotte Brontë, all of which provided her with useful material for her later novels, particularly *Shirley*. She received a further proposal of marriage from a curate in want of a wife, this time a Mr Pryce, and the charming William Weightman entered the Brontës' lives as their father's curate. There was a memorable visit to the sea at Bridlington for Charlotte and Ellen in late 1839, the arrival of young Martha Brown, the servant who was later to be the custodian of much Brontë memorabilia, and all the family unemployed and back home for Christmas that year. 1840 opened with Branwell's departure for yet another brief period of employment as a tutor to the Postlethwaite boys at Broughton-in-Furness, Cumbria. In May, Anne went to Thorp Green, where she was to remain for five years as governess with the Robinson family. Meanwhile, Charlotte immersed herself in reading French novels, developing a taste for the continent which was to emerge to strong and dramatic life-changing effect the following year. At the same time, she continued to write, as ever enjoying freedom from the yoke of work and responsibility. When Branwell returned from the Lake District, inspired in his writing by an encouraging encounter with Hartley Coleridge, Charlotte took advantage of the contact to send Coleridge some of her own work, a story about Caroline Vernon, the beautiful, illegitimate daughter of Northangerland and Louisa Vernon, falling gradually in love with the great Brontë hero, Zamorna. The correspondence does not survive, other than a highly flippant draft of a reply which seems unlikely to have been sent.

The need for employment was a perpetual problem. In October 1840, much to Charlotte's derision, Branwell took up a job on the Leeds and Manchester Railway in Calderdale. Then in March 1841 Charlotte became a private governess for the second and last time, with the Whites of Upperwood House, Rawdon. The struggle to reconcile herself to the governess's lot continued, but without success. Brilliant at acquiring knowledge, Charlotte was by then well aware that she had little talent for imparting it. By December 1841 she had left the post of governess once more, this time not entirely in despair, but with an exciting plan hatching in her mind.

~~~ *Brussels*

'Monsieur, the poor do not need a great deal to live on - they ask only the crumbs of bread which fall from the rich men's table – but if they are refused these crumbs - they die of hunger – No more do I need a great deal of affection from those I love – I would not know what to do with a whole and complete friendship – I am not accustomed to it – but you showed a little interest in me in days gone by when I was your pupil in Brussels – and I cling to the preservation of this little interest – I cling to it as I would cling to life.'

Charlotte Brontë to Constantin Heger, 8 January 1845

In 1913, nearly sixty years after her death, four love-letters written by Charlotte Brontë were published in *The Times*. The story of how they came to light, so long after they were written in 1844 and 1845, relates to the period of Charlotte's life

The Pensionnat Heger in Brussels where Charlotte was both student and teacher in 1842 and 1843

The Brontë Society

which many believe to have had the most important influence both on her writing and on her emotions. The love-lorn letters were written to Monsieur Constantin Heger, a married man, husband of her employer at the Pensionnat Heger in Brussels where Charlotte and Emily went in 1842 to improve their language skills, in order to enhance the prospects of the school the sisters hoped to open at Haworth Parsonage. Charlotte's Brussels years provided her with the material for two novels: *The Professor*, published posthumously in 1857, and *Villette*, published in 1853.

Charlotte's great friend Mary Taylor planned eventually to go to New Zealand to find a new life, no doubt encouraged by the propaganda of the day which encouraged single women to become emigrants. But first of all, in her quest for work and knowledge, she set off on a continental tour, and her letters describing European delights enthralled Charlotte. She wrote excitedly of Mary's experiences to Ellen on 7 August 1841:

> *'Mary's letter spoke of some of the pictures & cathedrals she had seen – pictures the most exquisite – & cathedrals the most venerable – I hardly know what swelled to my throat as I read her letter – such a vehement impatience of restraint and steady work. Such a wish for wings – wings such as wealth can furnish – such an urgent thirst to see – to know – to learn – something internal seemed to expand boldly for a minute – I was tantalised with the consciousness of faculties unexercised – then all collapsed and I despaired'.*

Charlotte had been thinking quite seriously of taking over Miss Wooler's school, now at Healds Hall, an idea which she had discussed in some detail in her correspondence with Ellen. Plans were so far advanced that she had asked Aunt Elizabeth Branwell for a loan of one hundred pounds in order to finance the project. However, Mary's letters from abroad flushed such thoughts from her mind, and in 1841 she rejected Miss Wooler's offer. Instead she became passionate about the idea of going to Europe. On 29 September 1841 Charlotte wrote to her aunt from her employers' home in Rawdon, asking if her offer of financial help could in part be put towards a study-trip to Brussels for herself and Emily:

'In half a year I could acquire a thorough familiarity with French. I could improve greatly in Italian, and even get a dash of German... Papa will perhaps think it a wild and ambitious scheme; but who ever rose in the world without ambition ? When he left Ireland to go to Cambridge University, he was as ambitious as I am now. I want us all to go on. I know we have talents, and I want them turned to account.'

Aunt Branwell agreed to the plan and from that moment on Charlotte was frantic with anticipation at the prospect of the trip to Belgium. All thoughts of the school in Dewsbury were cast to the winds – 'it is an obscure & dreary place ... in my secret soul I believe there is no cause to regret it' – and Charlotte's letters to her friend Ellen, from October 1841 until spring 1842 when she finally arrived in Brussels, were full of excited plans and speculation. Emily and Charlotte set off for

Arrival of a steamship; a mid-nineteenth-century scene.

The National Maritime Museum

Brussels on 8 February 1842, with their father to protect them on their first journey away from the north of England. Patrick went prepared with a home-made phrase-book of useful travellers' French, complete with guide to pronunciation – 'S'il vous plait montrez moi le priver = Sil voo play montray moa la prive – If you please shew me the privy'. They were accompanied by Mary Taylor and her brother Joe, who were once more making the crossing. Their base in London was the Chapter Coffee House in Paternoster Row in the heart of the city, which allowed Charlotte to satisfy the longings of a lifetime. With three days to spare before the packet sailed for Ostend she had time to explore the city of London, which until then she had seen only in her imagination. St Paul's, Westminster Abbey, the British Museum and the National Gallery were Charlotte's to enjoy at last. Her experience of the capital city is best summed up in the thoughts voiced by her heroine Lucy Snowe in *Villette*, who likewise savours the excitement of London on her way to the Channel ferry: 'I had a sudden feeling as if I, who had never yet truly lived, were at last about to taste life ... Prodigious was the amount of life I lived that morning.'

Ahead of the Brontës' party was a long voyage, which began early in the morning on Saturday 12 February with them boarding the Ostend packet at London Bridge Wharf. A fourteen-hour journey was followed by Sunday spent in Ostend, then a seventy-mile stagecoach ride to Brussels. The next day the Brontë sisters were introduced to the Pensionnat Heger in the Rue d'Isabelle, a plain building which hugged to itself the secret of a delightful walled garden. The garden was bordered by the 'Allée défendue', a gloomy, vine-covered walk which separated the Pensionnat from the Athenée Royal boys' school opposite. Both settings of the delightful garden and the mysterious alleyway were to feature to great effect in *The Professor* and *Villette*. The directrice was Madame Claire Zoë Heger, and her husband Constantin, who was a teacher at the next-door boys' school, gave lessons in literature to the girls. The school prospectus advertised a 'curriculum, founded on religious principles, basically comprises French language, History, arithmetic, geography and writing, as well as all the skills in needlework which a well-brought-up young lady requires.' Lessons in music and foreign languages were optional extras, and the fees for a year were 650 francs, which just about consumed the Brontës' budget.

In May, Charlotte wrote an entertaining letter to Ellen in which she gave revealing pen-portraits of Monsieur and Madame Heger:

> *'I was twenty-six years old a week or two since – and at that ripe time of life I am a school-girl – a complete school-girl and on the whole very happy in that capacity ... Madame Heger the head is a lady of precisely the same cast of mind and degree of cultivation & quality of character as Miss Catherine Wooler – I think the severe points are a little softened because she has not been disappointed & consequently soured – in a word – she is a married instead of a maiden lady ... Monsieur Heger the husband of Madame – he is professor of Rhetoric a man of power as to mind but very choleric & irritable in temperament – a little, black, ugly being with 'a face' that varies in expression, sometimes he borrows the lineaments of an insane Tom-cat – sometimes those of a delirious hyena – but very seldom he discards these perilous attractions and assumes an air not above a hundred degrees removed from what you would call mild & gentleman like ... Emily and he don't draw well together at all – when he is very ferocious with me I cry – & that sets things straight.'*

A page from one of Charlotte's Brussels exercise books, in which she developed both her knowledge of the French language and her writing skills, under the rigorous tuition of Monsieur Constantin Heger.

The Brontë Society

Monsieur Heger taught the girls French literature, and his excellence as a teacher and ruthlessness as a critic were as important for the development of Charlotte's writing as his own person was for the effect on her emotions. By dint of meticulous teaching he encouraged Charlotte to cut down on the flow of words, and taught her that her writing should be pared down into elegant shape. Heger's

Drawing of a landscape with fallen trees, which Charlotte produced during the summer vacation in Brussels in 1842. She and Emily both sketched the same woodland scene, on one of the few occasions when Charlotte actually studied from the life.

The Brontë Society

method was to present both sisters with essay assignments to be written in a particular literary style, having discussed a passage in class, exemplifying its good points. In that way they were introduced to serious French literature, they improved their understanding of the language, and they fine-tuned their writing skills. Examples of both Charlotte's and Emily's 'devoirs', or exercise books, survive today in the Brontë Parsonage Museum at Haworth and other libraries. The eagle-eyed teacher wrote copious comments in the margins, as on a piece of writing entitled *The Nest*, which Charlotte dated 30 April 1842, and which is now in the Berg collection in the New York Public Library

'*You must sacrifice*, without pity, *everything that does not contribute to clarity, verisimilitude and effect. Look with great suspicion on everything which sets off the main thought, so that the impression you give is highly coloured, graphic*'.

As their allotted six months in Brussels came to an end, the Hegers were persuaded by Charlotte to keep on the Brontës as pupils in exchange for teaching duties. Meanwhile, life back at home in Haworth was somewhat blighted. Branwell had been dismissed from his job on the railway because his lack of vigilance as chief clerk had allowed another employee to embezzle funds, the valley was disrupted by Chartist riots and, saddest of all, the much-loved local curate, William Weightman took ill with cholera and died on 6 September. His death was sharply followed by that of Elizabeth Branwell, the aunt who had brought up the young Brontës, and whose generosity had funded their trip to Brussels. She died on 29 October and was buried in the family vault beneath Haworth church before Charlotte and Emily could make the journey home. Just a few days earlier the Brontës suffered the sad loss of their young friend Martha Taylor from cholera, then a student at the Chateau Koekelberg, not far away from Brussels. So it was a melancholy return to Haworth on 8 November 1842, with the family reunited by death. The only light in their

This plan, drawn by Louise, daughter of Constantin and Zoe Heger, shows the layout of the school as it was in Charlotte's time there.

The Brontë Society

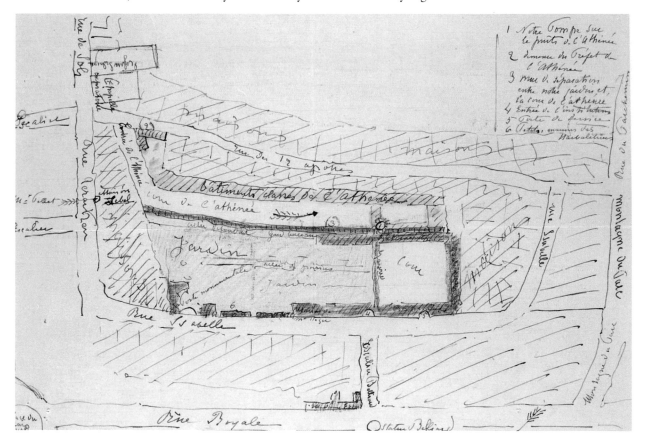

darkness was the arrival of a letter for Patrick Brontë from Constantin Heger which sang the Brontë sisters' praises:

> *'No doubt you will be pleased to hear that your children have made very notable progress in all the branches of instruction, and that this progress is entirely owing to their love of work and their perseverance; in dealing with such pupils we have had but little to do; their progress is your handiwork much more than ours; we have not had to teach them the value of time and instruction, they had learnt all that in their father's house, and we for our part have had the minor merit of guiding their efforts and providing suitable material to foster the admirable activity that your daughters have derived from your example and your lessons.'*

Heger goes on to suggest that for the future one or both of the Brontës should return to Brussels to teach in the school – an idea that thrilled Charlotte as much as it appalled Emily. For the time being though, Charlotte was tied to Haworth and mourning for her aunt. Elizabeth Branwell's death brought with it a new measure of independence for the Brontë women as each, along with their Penzance cousin Eliza Kingston, inherited an equal share of her estate, amounting to just under three hundred pounds, most of it invested in shares in the York and North Midland Railway. Christmas 1842, then, found Emily relieved to be at home and happy to adopt the role of housekeeper once more, Anne settled in her post at Thorp Green, planning to return with brother Branwell as tutor to Edmund Robinson, and Charlotte intent on travelling back alone to Belgium as soon as she decently could. In fact, 27 January 1843 found her setting off on the nine o'clock train from Leeds to London, and from there to Brussels where she arrived at the Pensionnat Heger in her new role as a teacher, 'Mademoiselle Charlotte'.

Loneliness came unexpectedly to Charlotte on her return to the school. She had not reckoned on how much she would miss Emily's company, Mary Taylor had left to teach in Germany and poor Martha Taylor was no more. Too timid to spend her evenings in the Heger family's sitting-room, as invited, Charlotte began to depend more and more on the pleasure she found in Constantin Heger's company when they worked together, he teaching her French, and she teaching English to Heger and his brother-in-law – 'if you could see and hear the efforts I make to teach them to pronounce like Englishmen and their unavailing attempts to imitate, you would laugh to all eternity.' She also concentrated on her German studies, taking expensive private lessons for which she translated German poetry into French as well as English. In her increasingly precious time with Monsieur Heger, Charlotte began to work at translating English poetry into French and vice versa. She revelled in the fact that Heger clearly regarded her as an outstanding pupil, and for the first time she enjoyed an intellectually equal relationship with a man outside her own family circle, one with whom she could argue about the nature of genius, and one to whom she felt increasingly able to write about her personal feelings, under the guise of literary criticism. Very soon in 1843, Charlotte's relationship with this older, married man had evolved into the prototype for what was to become one of the most celebrated pairings in English literature, that of the plain little governess Jane Eyre and her volatile employer, Mr Rochester. Charlotte's respect for Heger had turned into obsessive love.

The rest of Charlotte's Brussels existence was becoming tiresome to her. The people were dull, the food poor, the weather horribly wet, and Madame Heger insufferably cold-hearted. The long summer vacation filled her with dread, and in the event she was driven to wandering the streets of the city, even finding herself one day in the Catholic cathedral of Sainte Gudule. In a letter to Emily Brontë she described what happened that day, an incident that was later to find its way into her great Brussels novel, *Villette*:

> '*I took a fancy to change myself into a Catholic and go and make a real confession to see what it was like... The priest asked if I was a Protestant then. I somehow could not tell a lie, and said 'yes'. He replied that in that case I could not 'jouir du bonheur de la confesse'; but I was determined to confess, and at last he said he would allow me because it might be the first step towards returning to the true church. I actually did confess – a real confession.*'

The Misses Brontë's Establishment

FOR

THE BOARD AND EDUCATION

OF A LIMITED NUMBER OF

YOUNG LADIES,

THE PARSONAGE, HAWORTH,

NEAR BRADFORD.

Terms.

	£.	s.	d.
BOARD AND EDUCATION, including Writing, Arithmetic, History, Grammar, Geography, and Needle Work, per Annum,	35	0	0
French, German, Latin — each per Quarter,	1	1	0
Music, Drawing, — each per Quarter,	1	1	0
Use of Piano Forte, per Quarter,	0	5	0
Washing, per Quarter,	0	15	0

Each Young Lady to be provided with One Pair of Sheets, Pillow Cases, Four Towels, a Dessert and Tea-spoon.

A Quarter's Notice, or a Quarter's Board, is required previous to the Removal of a Pupil.

Was it the burden of her growing feelings for a married man that had caused Charlotte Brontë, with her loathing of Catholicism, to turn to the confession-box in a Roman Catholic cathedral in a foreign land ?

By October Charlotte was covertly trying to persuade her family to provide her with a reason to return to Haworth. In a moment of utter misery she was moved to write on the inside cover of her atlas: 'I am very cold – there is no fire – I wish I were at home with Papa – Branwell – Emily – Anne & Tabby – I am tired of being amongst foreigners it is a dreary life – especially as there is only one person in this house worthy of being liked – also another who seems a rosy sugar-plum but I know her to be coloured chalk.' Two months later she wrote to Emily to tell her family she was in very low spirits and on her way home. On 23 January 1844 she shared a little of her misery with Ellen in an introspective letter in which she dwells on her past – 'I suffered much before I left Brussels – I think however long I live I shall not forget what the parting with Monsr Heger cost me – It grieved me so much to grieve him who has been so true and kind and disinterested a friend' – and her future, 'something in me which used to be enthusiasm is tamed down and broken – I have fewer illusions – what I wish for now is active exertion – a stake in life – Haworth seems such a lonely, quiet spot, buried away from the world'. In fact the only scheme in view for the future was the old plan to open the Brontës' own school for girls at the Parsonage. However, the rather desultory attempts to advertise the school met with rapid failure and the plan was finally abandoned for good in the summer of 1844. Charlotte conveyed little disappointment about this demise; her thoughts were

still with the object of her affections in Brussels, and in July and October that year, and again in January 1845, Charlotte penned the passionate letters which were revealed to a startled literary world in the pages of *The Times* on 29 July 1913.

The Heger letters in the British Library in London make a poignant sight. Ripped into pieces by Constantin Heger, glued and sewn back together again by Madame Heger, who retrieved them from the waste-bin and kept them, should evidence ever be needed that Charlotte Brontë was not blameless in her behaviour towards the lady's husband. Now preserved between heavy sheets of glass like flies in amber, the letters encapsulate the small tragedy of Charlotte Brontë's unrequited love. Before the revelation of 1913, scholars had identified the Brussels years as marking a turning point in Charlotte's life. Indeed, Elizabeth Gaskell had been shown the letters by Heger in 1856, but she had chosen, as she had done with other material, to suppress their significance. Again in 1894, a researcher shown the letters by Heger's daughter Louise, also chose to conceal their existence from the rest of the world. The four surviving letters were in fact just a few of many. Apparently the besotted Charlotte had for a time written as often as twice a week. Finally, in 1913, Heger's son Paul and his sisters offered the letters to the British Museum so that literary scholars of the future could make up their own minds about the nature of the relationship between Charlotte Brontë and her Belgian teacher. The saddest fact of all is that none of Charlotte Brontë's love-letters ever received a loving reply.

Opposite page:

Constantin Heger and his family in a portrait painted within a few years of Charlotte's departure from Brussels. Charlotte described Heger, the object of her unrequited passion, as 'the black swan', and he formed the inspiration for the characters of Edward Rochester in Jane Eyre *and Monsieur Paul Emmanuel in* Villette.

The Brontë Society

～ *Poems*

Throughout most of the year 1845, Charlotte was deep in the slough of despond, struggling with her turbulent feelings for Monsieur Heger and, as she approached thirty, bemoaning the dullness of her life and her lack of accomplishment. On 24 March, the month that Mary Taylor finally left on her adventurous trip to New Zealand, Charlotte, full of remorse for the loss of her friend to distant lands, wrote to Ellen:

'I can hardly tell you how time gets on here in Haworth – there is no event whatever to mark its progress – one day resembles another -and all have heavy lifeless physiognomies – Sunday – baking day & Saturday are the only ones that bear the slightest distinctive mark – meantime life wears away – I shall soon be 30 – and I have done nothing yet – Sometimes I get melancholy – at the prospects before and behind me – yet it is wrong and foolish to repine – undoubtedly my Duty directs me to stay at home for the present – There was a time when Haworth was a very pleasant place to me, it is not so now I feel as if we were all buried here'.

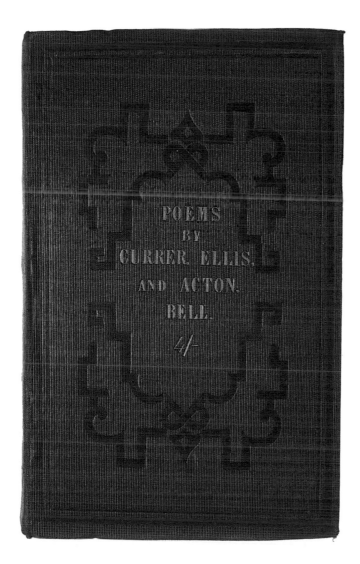

When it was first published in 1846, the Poems of Currer, Ellis and Acton Bell sold just two copies. First editions of the book are today highly sought-after collectors' items.

The Brontë Society

It was also a year of crisis in the lives of other members of her family. There was general concern for Patrick Brontë whose sight was failing because of cataracts on his eyes. The new young Irish curate, Arthur Bell Nicholls, who arrived in Haworth in May 1845, was at once a great help to the disabled Reverend Brontë.

Charlotte's depression stemmed not only from her emotional misery, but also from her preoccupation with money and security, yet she displayed a certain indolence by staying at home and not actively looking for a job. Then in June, about the time that Anne Brontë resigned from her post with the Robinsons of Thorp Green, the significance of her departure at that point unknown, Charlotte's darkness was temporarily lightened by a visit to Hathersage in Derbyshire with Ellen Nussey. Ellen's brother Henry, once a suitor of Charlotte herself, had now married and been appointed curate at Hathersage. The three-week visit, as well as cheering Charlotte's spirits, also proved to be a major influence on the shaping of her novel, *Jane Eyre*. As Charlotte enjoyed her visits around the neighbourhood, her writer's eye took in the village and its setting, the Eyre family memorial brasses in the local church, North Lees Hall, formerly the home of the Eyre family and the flat moorland landscape of Derbyshire. Within a short time these locations and details were to be deployed in the writing of a novel which was to change Charlotte's life forever.

The homecoming from the delightful holiday in July was not so uplifting. Brother Branwell was found to be at Haworth, sacked from his post as tutor with the Robinson family and languishing at home. Unknown to him, he was suffering from the same malady as his older sister, a passion for a married person. Branwell had apparently been dismissed from Thorp Green having been discovered in an affair with Lydia Robinson, the lady of the house. On 31 July, Charlotte wrote in unsympathetic tones of her unfortunate brother's state:

> '*I found Branwell ill – he is so very often owing to his own fault - I was not therefore shocked at first – but when Anne informed me of the immediate cause of his present illness I was greatly shocked... We have had sad work with Branwell since – he thought of nothing but stunning, or drowning his distress of mind...*'

Branwell's demise had begun, but even in his distress he turned his mind to the business of earning a living, and announced his intention of writing a novel. This had an effect on Charlotte, and she too began to consider how the family compulsion to write could be turned to money-earning effect. In her *Biographical Notice* about her sisters, published with *Wuthering Heights* in 1850, Charlotte describes how her idea for the Brontë sisters' first publishing venture was born:

> *'One day, in the autumn of 1845, I accidentally lighted on a MS volume of verse in my sister Emily's handwriting. Of course, I was not surprised, knowing that she could and did write verse; I looked it over, and something more than surprise seized me, – a deep conviction that these were not common effusions, nor at all like the poetry women generally write. I thought them condensed and terse, vigorous and genuine. To my ear, they had also a peculiar music – wild, melancholy and elevating.'*

As she secretly read her sister Emily's poems, Charlotte at last saw an opportunity for making their writing work: the sisters would publish jointly a volume of their poetry. Emily took some reconciling to the idea, and Charlotte related how first of all it took her many hours to make amends for having had the audacity to read Emily's poems without permission, and days to persuade her that such poems merited publication. Anne, according to Charlotte, 'quietly produced some of her own compositions' to add to the project. Clearly, it was the ambitious Charlotte who was the driving force behind the move to publish the poems. All three sisters agreed on one thing, the need to use pseudonyms:

> *'Averse to personal publicity, we veiled our own names under those of Currer, Ellis and Acton Bell; the ambiguous choice being dictated by a sort of conscientious scruple at assuming Christian names positively masculine, while we did not like to declare ourselves women, because – without at that time suspecting that our mode of writing and thinking was not what is called 'feminine' – we had a vague impression that authoresses are liable to be looked on with prejudice.'*

*Charlotte's first letter to
Aylott & Jones, London:
'Gentlemen, May I
request to be informed
whether you would
undertake the publication
of a collection of short
poems in I vol. oct-
If you object to publishing
the work at your own risk
– would you undertake it
on the Author's account?
– I am Gentlemen,
Your obdt. hmble. Servt.,
C. Brontë'*

The Brontë Society

Possessed once more by determination and the satisfaction that came with hard work, Charlotte and her sisters spent the autumn selecting the poems for their slim volume. For Charlotte, her enthusiasm for the enterprise was more to do with the business of getting published than for her own actual contribution to the volume, for she contributed the smallest number of poems, nineteen in all, none of them recent and some of them dating back as far as 1837. Anne contributed twenty-one poems, all of them written since 1840, and Emily also twenty-one, two from 1839, but the vast majority very recent. Emily insisted, wisely, on editing out the mystifying Gondal references, and she supplied the best known poem in the volume:

'No coward soul is mine
No trembler in the world's storm-troubled sphere
I see heaven's glories shine
And Faith shines equal arming me from fear'

Producing the *Poems* acted as a means of boosting Charlotte out of her post-Brussels torpor. She began to write to prospective publishers, asking them their terms. The correspondence with Aylott and Jones began on January 28 1846 and rapidly gathered pace. The cost having been ascertained, the poems were then sent off in two separate parcels, over which Charlotte fretted for their safe arrival. This was quickly followed by a banker's draft for £31 10s to cover the cost of the publication, money that came out of the legacy left to the sisters by their aunt. Happy that all was going well, Charlotte then broke her self-enforced stay in Haworth and went to visit Ellen, during which time she took care to consult the surgeon husband of Ellen's cousin about her father's eyesight, to learn that an operation as soon as possible was the best course of action.

[Facsimile of handwritten letter]

'Sir, My relatives, Ellis and Acton Bell and myself, heedless of the repeated warnings of various respectable publishers, have committed the rash act of printing a volume of poems. The consequences predicted have, of course, overtaken us; our book is found to be a drug; no man needs or heeds it; in the space of a year our publisher has disposed but of two copies, and by what painful efforts he succeeded in getting rid of these two, himself only knows. Before transferring the edition to the trunk-makers, we have decided on distributing as presents a few copies of what we cannot sell — we beg to offer you one in acknowledgement of the pleasure and profit we have often and long derived from your works — I am Sir, Yours very respectfully, Currer Bell'

Charlotte Brontë to J. G. Lockhart, 16 June 1847

The Brontë Society

Back home, she immersed herself in proof-reading the poems, an activity that was kept as secret as possible from the rest of the household. It is notable that Branwell, already in fact a published poet, his verse having appeared in the *Halifax Guardian*, was not invited to take part in the venture.

May 7 1846 must have been an exciting day at Haworth Parsonage, for it was then that the first three copies of the *Poems* arrived. Consisting of one-hundred-and-sixty-five pages, price four shillings, how thrilling the volume must have appeared to Charlotte's eyes, no doubt summoning up remembrance of the young Brontës' first crude, literary efforts, the childhood little books, hand-crafted from scraps of household paper. The ambitious Charlotte at once urged the publishers to send out review copies, but she had to wait two months to read anything of the work's reception from the critics. Furthermore, later that month the family was plunged into gloom by Branwell's desperate reaction to the news that Mr Robinson of Thorp Green had died, yet Lydia had no wish to see him, and fed him the tale that her husband's will specifically forbade her to marry Branwell, otherwise she would lose her fortune.

The Brontës were perturbed to discover that their first review, in the *Critic*, was interested in the identity of the Bells - 'Who are Currer, Ellis, and Acton Bell, we are nowhere informed', an issue that was to dog them for the rest of their lives. In the *Athenaeum*, Sydney Dobell picked out Emily's poems for special merit – 'A fine quaint spirit has the latter, which may have things to speak that men will be glad to hear, – and an evident power of wing that may reach heights not here attempted.' But the *Critic*, gratifyingly, was very fulsome with praise all round: 'here we have good, wholesome, refreshing, vigorous poetry – no sickly affectations, no namby-pamby, no tedious imitations of familiar strains, but original thoughts, expressed in the true language of poetry'. Charlotte was thrilled and ordered the publishers to spend ten pounds on advertisements quoting this review.

It could be said that the *Poems* of Currer, Ellis and Acton Bell was one of the most celebrated publishing failures of the nineteenth century. Two copies were sold. Undeterred, Charlotte briskly sent off some of the embarrassing surfeit of copies as presentation volumes to the great literary figures of the day, to J.G. Lockhart, Hartley Coleridge, William Wordsworth, Alfred Lord Tennyson, Thomas De Quincey and Ebenezer Elliott. She had enjoyed the whole process of getting the poems to press, and she had learnt from it: poems were not as popular as novels, and vanity publishing was a very expensive business. The failure of the poems was perhaps not such a worry because Charlotte, Emily and Anne were already on a new literary path. As early as 6 April 1846, Charlotte had written to Aylott and Jones, telling them that the Bells 'are now preparing for the Press a work of fiction – consisting of three distinct and unconnected tales which may be published either together as a work of 3 vols. of the ordinary novel-size, or separately as single vols.'

~ *Jane Eyre*

'Do you think because I am poor, obscure, plain and little, I am soulless and
heartless? – You think wrong! – I have as much soul as you, – and full as much
heart! And if God had gifted me with some beauty, and much wealth, I should
have made it as hard for you to leave me, as it is now for me to leave you. I
am not talking to you now through the medium of custom, conventionalities, nor
even of mortal flesh: – it is my spirit that addresses your spirit; just as if both had
passed through the grave, and we stood at God's feet, equal, – as we are!'

Jane Eyre, *Chapter 23*

J ane Eyre is an orphaned child, left to the loveless care of her aunt, Mrs Reed,
whose harsh treatment of Jane provokes a passionate outburst which condemns
her to Lowood, a miserable charitable school. Jane remains at Lowood as pupil and
teacher until she makes her break into the world and goes as governess to Adele, the
illegitimate daughter of Edward Rochester, the sardonic and darkly attractive owner
of Thornfield Hall. Despite her lack of beauty, Jane's brave spirit and sharp wit
captivates Rochester, and they fall in love. Their marriage is stopped at the altar by
the discovery of Rochester's mad wife, Bertha, who is incarcerated in the attic at
Thornfield. Jane flees, and is rescued by the Rivers family. The Reverend St John
Rivers and his sisters nurse her back to health, and St John Rivers asks Jane to go
with him as his wife on his mission to India. But an eerie call summons Jane back to
Thornfield to find that the Hall has been burnt to the ground and that Rochester has
been blinded and injured in an attempt to save his wife, who has perished. Jane finds
Rochester again, lost in his despair. They marry and together find true happiness.

Charlotte Brontë's *Jane Eyre* is a classic novel in the truest sense of the word.
From the time of its first publication in October 1847, at once a popular success, it
has never been out of print. It has been translated into countless languages and is
perpetually rediscovered by new generations of readers. The powerful tale of a little
governess 'disconnected, poor, and plain' has caught the imagination of millions.
Jane Eyre overcomes both the disadvantages of her unhappy, orphan background and

Jane Eyre

by ~~Currer Bell~~

Vol. 1st

Chap. 1st

There was no possibility of taking a walk that day.
We had been wandering indeed in the leafless shrubbery
an hour in the morning, but since dinner (Mrs Reed
when there was no company, dined early) the cold win
wind had brought with it clouds so sombre, a rain so
etrating that further out-door exercise was now out of
question.

I was glad of it; I never liked long walks — especia
on chilly afternoons; dreadful to me, was the coming hom
in the raw twilight with nipped fingers and toes and a
saddened by the chidings of Bessie, the nurse, and humble
by the consciousness of my physical inferiority to Eliza, Joh
and Georgiana Reed.

The said Eliza, John and Georgiana were now clustere
round their Mamma in the drawing-room; she lay rec

the dramas that beset her when she goes to work at Thornfield Hall for the fearsome and fascinating Mr Rochester, to find a sense of self and a marriage of true love. The book has affected later writers to such a degree that other well-known and much-loved novels have taken up the theme of the loveless girl and her pursuit of happiness, Daphne Du Maurier's *Rebecca* and Frances Hodgson-Burnett's *The Secret Garden* to name but two. In the latter part of the twentieth century, Jane Eyre has come to be seen as a feminist heroine and her character has been closely identified with that of its creator. It is certainly true that Charlotte set out purposely to create a heroine who was not beautiful, rich and accomplished, but a woman who is small, plain, penniless and alone in the world. Her intention was to some extent to turn romantic fiction on its head, to have her female character wrestle with the moral issues central to her relationship with the man she loves. Jane Eyre, as she tells Rochester on more than one occasion, has a brain in her head, and her own intellect speaks to her as clearly as her heart. When Jane and Rochester are finally united in marriage, theirs is an equal partnership. Conveniently, Jane has inherited a fortune from her deceased uncle in the West Indies, another issue which was regarded by the writer as perhaps the most important key to women's independence, financial self-sufficiency. Although many readers relate to the book as a straightforward love story its great power is also attributable to its complexity as a statement of the position of women in the mid-nineteenth-century.

Jane Eyre was Charlotte's first published novel, but it was not the first book she wrote. When she wrote tantalisingly to Aylott and Jones of 'three distinct and unconnected tales' by the mysterious Bell brothers, she referred to Emily's *Wuthering Heights*, Anne's *Agnes Grey* and her own novel *The Professor*, which tells the story of William Crimsworth, a young man who goes to seek his fortune as a schoolteacher in Brussels and falls in love with Frances Henri, an Anglo-Swiss pupil-teacher. *The Professor* is of course based on Charlotte's Brussels experience, but is far less successful than her later autobiographical novel *Villette*. But writing her first novel formed a very important part of Charlotte's literary apprenticeship, a course that had begun back in her childhood with the colourful tales of Angria and Glasstown, through her journals and letters, writing her own verse and at last using her own life experiences as the inspiration for her art.

Opposite page:

The manuscript of Jane Eyre *was bequeathed to the British Museum in 1914 by Mrs Elizabeth Smith as a memorial of her late husband, George Smith, Charlotte's publisher, who died in 1901. It is the autograph fair copy in three volumes which Charlotte sent to Smith, Elder and Co. The first page of the manuscript contains one of the most famous opening sentences in English literature: 'There was no possibility of a walk that day.'*

The British Library Add. MS 43474, f.20

By 27 June 1846, Charlotte had completed the fair copy of *The Professor*. The Brontës' first three novels had been written in close collaboration as in childhood, with the nocturnal walks around the dining-room table as they talked and criticised and shared their ideas. In writing *The Professor*, Charlotte had not quite made the break from her Angrian Gothic manner. The very fact that she wrote in the male first person is a hang-over from the days of Angria, where it was always the heroes who had the best adventures. It was as if she had not quite grown up in her writing. Set alongside Anne Brontë's heart-rendingly realistic tale of the trials of a governess, *Agnes Grey*, and Emily Brontë's staggering, poetic tour-de-force, *Wuthering Heights*, Charlotte's first novel was the weakest. From July 1846, the parcel of the three Brontës' first novels set off on a weary trail round the publishers that was to last a year, ending in a triumph of a sort for Emily and Anne when Thomas Cautley Newby took the books onto his list, because once again they had to agree to pay for the printing themselves, and disappointment for Charlotte when *The Professor* was rejected. But much was to happen for its author in the meantime.

As the sisters packed up their parcel of novels in July 1846, their greatest family concern was with their father. By then, Patrick was almost blind and an operation had become essential. In August, Charlotte and the Reverend Brontë left Haworth for Manchester, where the sisters had located an eye-surgeon. It was in the heart of red-brick Manchester in lodgings off the Oxford Road, where, slightly distracted by fretting over the housekeeping arrangements as her father recuperated in a darkened room, Charlotte Brontë began to write her greatest novel. She later told Harriet Martineau how she had rapidly become a woman possessed, writing feverishly as she poured out the story. For the first time in her life Charlotte was putting her writing skills to work hand-in-hand with her feelings. Into the opening chapters she poured all her grief and rage at the injustices of the place where she lost her two older sisters, Maria and Elizabeth, over twenty years before. Cowan Bridge school was recreated as Lowood, set in a dank landscape with creeping fogs, ruled by the cold Mr Brocklehurst. The proud and defiant Jane Eyre was born. Charlotte called upon her writer's memory of places, people, names and anecdotes, interweaving features of her own life with her fictional creation. The congenial visit to Derbyshire with Ellen was recalled to provide the landscape through which her

heroine journeyed. Local legends of mad wives incarcerated in remote houses intermixed with Charlotte's encyclopaedic knowledge of myth and fairy tale, so that the corridors of Thornfield recall those of Bluebeard's palace. Rochester – Byron, Heger, Zamorna in one – rides into Jane's life heralded by a great dog, like the terrifying gytrash of darkest childhood terrors. On she wrote, until her heroine had lived through the ravaging history of falling in love with her employer, finding her way to the altar to meet him, only to be rent apart from him by the awful revelation of the mad wife in the attic.

Charlotte carried on writing back home at Haworth, her enthusiasm dimmed a little by the constant rejection of her manuscript of *The Professor*. On July 15 1847 she decided to have one more attempt to sell the book, and sent it to Smith, Elder & Co. of 65, Cornhill, London, not even bothering to repack the bundle in new wrapping-paper, but sending it with the last publisher's address crossed out and the new one written in its place. Not an encouraging sight for the recipients, but nonetheless it was a move which was to make literary history, for Charlotte was delighted to receive a letter from Smith, Elder & Co. which, although they declined to publish *The Professor*, contained careful criticism of her script and declared an

> vision from another world – my heart beat thick – my head grew hot – a sound filled my ears which I deemed the rushing of wings – something seemed near me, I was oppressed – suffocated, endurance broke down – I uttered a wild, involuntary cry – I rushed to the door and shook the lock in desperate effort. Steps came running along the outer passage; the key turned, Bessie and Abbot entered.
>
> "Miss Eyre, are you ill?" said Bessie.
>
> "What a dreadful noise! It went quite through me" exclaimed Abbot.
>
> "Take me out! Let me go into the nursery!" was my cry.
>
> "What for? Are you hurt? Have you seen something?" Again demanded Bessie.
>
> "Oh! I saw a light – and I thought a ghost would come". I had now got hold of Bessie's hand, and she did not snatch it from me.
>
> "She has screamed out on purpose;" declared Abbot in some disgust "And what a scream! If she had been in great pain one would have excused it, but she only wanted to bring us all here; I know her naughty tricks."
>
> "What is all this?" Demanded another voice peremptorily, and Mrs Reed came along the corridor, her cap flying wide, her gown rustling stormily. "Abbot and Bessie, I believe I gave orders that Jane Eyre should be left in the red-room till I came

interest in her next novel. Spurred on by this, Charlotte raced to finish *Jane Eyre*, and within two weeks, on 24 August 1847, her novel was in the post.

There then followed an anxious two weeks whilst the publishers considered. In fact, the readers were united in their view that *Jane Eyre* was a masterpiece. George Smith, owner of the firm, was intrigued and elected to read the manuscript

"I believe – I have faith: I am going to God"

"Where is God? What is God?"

"My Maker and yours, who will never destroy what he created; I rely implicitly on his power, and confide wholly in his goodness; I count the hours till that eventful one arrives which shall restore me to him – reveal him to me."

"You are sure then, Helen, that there is such a place as heaven, and that our souls can get to it when we die?"

"I am sure there is a future state – I believe God is good – I can resign my immortal part to him without any misgiving – God is my father – God is my friend – I love him – I believe he loves me."

"And shall I see you again, Helen, when I die?"

"You will come to the same region of happiness – be received by the same mighty, universal Parent – no doubt, dear Jane."

Again I questioned – but this time only in thought – "Where is that region? Does it exist?" And I clasped my arms closer round Helen; she seemed dearer to me than ever; I felt as if I could not let her go; I lay with my face hidden on her neck. Presently she said in the sweetest tone:

"How comfortable I am! That last fit of coughing has tired me a little – I feel as if I could sleep; but don't leave me, Jane, I like to have you near me."

"I'll stay with you, dear Helen; no one shall take me away."

134

Charlotte's description of the death of Helen Burns is moving in the extreme:
' "And shall I see you again, Helen, when I die?" "You will come to the same region of happiness – be received by the same mighty, universal Parent – no doubt, dear Jane." Again I questioned – but this time only in thought - "Where is that region? Does it exist?" And I clasped my arms closer round Helen; she seemed dearer to me than ever; I felt as if I could not let her go; I lay with my face hidden on her neck'.

The British Library
Add. MS 43474, f.134

for himself. Mrs Gaskell, in her *Life* of Charlotte related his account of the effect the book had on him:

'*After breakfast on Sunday morning I took the MS of 'Jane Eyre' to my little study, and began to read it. The story quickly took me captive. Before twelve o'clock my horse came to the door, but I could not put the book down...[I] went on reading the MS. Presently the servant came to tell me that luncheon was*

Hundreds of Charlotte Brontë's letters survive, but this one sent to Smith, Elder and Co. on 24 August 1847 is perhaps one of the best known:

'Gentlemen, I now send you per Rail - a M.S. entitled "Jane Eyre, a novel, in 3 vols. by Currer Bell". I find I cannot pre-pay the carriage of the parcel as money for that purpose is not received at the small Station-house where it is left. If, when you acknowledge the receipt of the M.S. you will have the goodness to mention the amount charged on delivery - I will immediately transmit it in postage stamps. It is better in future to address Mr. Currer Bell - under cover to Miss Brontë - Haworth - Bradford Yorks - as there is a risk of letters otherwise directed, not reaching me at present - To save trouble I enclose an envelope- I am Gentlemen Yours respectfully C Bell'.

Private Collection

ready; I asked him to bring me a sandwich and a glass of wine, and still went on with "Jane Eyre". Dinner came; for me the meal was a very hasty one, and before I went to bed that night I had finished reading the manuscript.'

Despite their great enthusiasm for *Jane Eyre*, the publishers nonetheless struck a hard bargain, which, although she made some objections to the one hundred pounds offered and positively refused to rework the text, Charlotte quickly accepted. This accomplished, events then moved at a great pace. Within a week she had received the proofs, corrected and returned them. Smith, Elder & Co. were far more

'*Jane Eyre encounters
Rochester.*'
Illustration to Jane Eyre
*by Fritz Eichenberg
Random House edition
1943*

The Brontë Society

Jane Eyre's first meeting with Rochester is highly dramatic, and from the moment that he literally rides into her life and crashes at her feet, heralded by the dog Pilot, the dull life of the little governess is transformed: '…they had slipped on the sheet of ice which glazed the causeway. The dog came bounding back, and seeing his master in a predicament, and hearing the horse groan, barked till the evening hills echoed to the sound which was deep in proportion to his magnitude.'

The British Library
Add. MS 43474, f.188

188

horse were down; they had slipped on the sheet of ice which glazed the causeway. The dog came bounding back, and seeing his master in a predicament, and hearing the horse groan, barked till the evening hills echoed the sound which was deep in proportion to his magnitude. He snuffed round the prostrate group and then he ran up to me; it was all he could do; there was no other help at hand to summon. I obeyed him and walked down to the traveller, by this time struggling himself free of his steed. His efforts were so vigorous, I thought he could not be much hurt; however I asked him the question:

"Are you injured, sir?"

I think he was swearing but am not certain, however he was pronouncing some formula which prevented him from replying to me directly.

"Can I do anything?" I asked again.

"You must just stand on one side" he answered as he rose, first to his knees and then to his feet. I did, whereupon began a heaving, stamping, clattering process, accompanied by a barking and baying which removed me effectually some yards' distance; but I would not be driven quite away till I saw the event. This was finally fortunate; the horse was reestablished and the dog was silenced with a "Down Pilot!". The traveller now stooping, touched his foot and leg, as if trying whether they were sound; apparently something ailed them, for he halted to the stile

efficient publishers than T C Newby. *Wuthering Heights* and *Agnes Grey* had still not reached the press when, on the morning of 19 October 1847, Charlotte received the first six copies of *Jane Eyre*, printed in three volumes bound with cloth covers. She expressed to the publishers her pleasure with the quality of production, then braced herself for the reviews. She was not, at first, to be disappointed. William Makepeace Thackeray, Charlotte's greatest literary hero, had been sent a review copy of *Jane Eyre* and his reply was forwarded to Charlotte by William Smith Williams, her confidant at Smith, Elder & Co. It contained words of praise that Charlotte had longed for all of her life, and from such an illustrious source:

The existence of the mad wife in the attic is introduced to the reader when Jane hears screams and evil laughter: 'This was a demoniac laugh, low, suppressed and deep, muttered, as it seemed at the very key-hole of my chamber-door. The head of my bed was near the door and I thought at first, the goblin-laughter stood at my bedside, or rather, crouched by my pillow...'

The British Library
Add. MS 43474, f.259

'I wish you had not sent me Jane Eyre. It interested me so much that I have lost (or won if you like) a whole day in reading it at the busiest period, with the printers I know waiting for copy. Who the author can be I can't guess – if a woman she knows her language better than most ladies do, or has had a 'classical' education. It is a fine book though – the man and woman capital – the style very generous and upright so to speak...I have been exceedingly moved & pleased by Jane Eyre. It is a woman's writing, but whose? Give my respects and thanks to the author – whose novel is the first English one (& the French are only romances now) that I've been able to read for many a day.'

The *Critic*, the *Era* and *The Examiner* were abundant in their praise, although the *Spectator* panicked Charlotte by condemning what they called the low tone of behaviour in the book. But the book's popular success was indisputable. The first edition of about two and a half thousand copies sold out within three months, and *Jane Eyre* was reprinted again in January and April 1848. For the second edition, Charlotte persuaded her publishers to let her insert a preface, because she wanted to answer the whispers of impropriety from her critics, in which she expressed the view that, 'Conventionality is not morality. Self-righteousness is not religion.' She was also allowed to dedicate the edition to Thackeray, a move that turned out to be a blunder. The sad truth was that Thackeray's own wife had become insane after just four years of marriage and was committed to an asylum. Furthermore, in Thackeray's novel *Vanity Fair*, Becky Sharp is also a governess who marries her employer.

London began to buzz with the gossip that Currer Bell had been governess in the Thackeray household. Charlotte was mortified when she found out that her compliment had been turned upside down.

As the year 1847 drew to a close, the Brontë sisters at last found themselves to be published authors, earning something of a living from their writing. Anne's *Agnes Grey* and Emily's *Wuthering Heights* were finally published in December, to their chagrin still full of the typographical errors which they had laboured to correct. Their father Patrick was active once more, his sight repaired by the operation in the autumn. Branwell, sadly, gave cause for concern to all the family, as he slid still further into an abyss of his own making. An indication, had Charlotte but known it, of the great grief that was to mar her new-found and long cherished literary success in the months to come.

⬲ *Deaths*

Charlotte's relationship with her brother Branwell, although loving, was always competitive, and as the years went by the once close siblings drifted apart. From earliest childhood, the two vied with each other to dominate their shared fantasy world from which the juvenile writings sprang. As fast as one of them invented a daring hero or glamorous heroine, the other struck back with an equally stirring creation. Branwell had the privilege of remaining at home to be taught by his scholarly father. Whilst Charlotte was away at school, she worried that Branwell was busy killing off her best characters in her absence. As the Brontës grew up, and Charlotte struggled with the inequities of a woman's lot, especially the need to earn a living by dreary governessing, it must have been galling in the extreme to witness the advantages doled out to Branwell, because he was male. It was she who slaved at her drawings, perfecting her artistic techniques as best she could, cherishing the ambition to be an artist. It was Charlotte who won the honour of exhibiting two of her meticulous drawings at the prestigious Leeds exhibition in 1834. Yet it was

Previous pages:

A view of Bradford in the early stages of its industrial development, painted by John Wilson Anderson (1792–1851) between 1825 and 1833. Anderson was a painter of houses and stage scenery as well as pictures, and a member of the same intemperate Bradford artistic circle as Branwell Brontë

Bradford Art Galleries and Museums

Branwell who was given expensive lessons in painting in oils by the Leeds portraitist, William Robinson. As the boy of the family, Branwell had far more career options open to him, and at first it seems to have been assumed by all around him that his natural talents would place the glittering prizes of life easily within his grasp.

The first disappointment came when he was eighteen years old, and the much vaunted plan to apply to the Schools of the Royal Academy of Arts came to naught. In fact, Branwell did not even get as far as making the trip to London, as was previously believed. For several years he pottered around in village life, painting portraits of his father's wealthier parishioners and local tradesmen, his career as artist culminating in a meagre twelve months as a portrait painter in Bradford in 1838–1839. Surviving letters written by Patrick Brontë, enquiring of business acquaintances about employment for his son, show that by the time Branwell was twenty, his inability to settle down to a career was a cause for fatherly concern.

In the last eight years of his life, Branwell, like Charlotte, was employed infrequently, only in his case each post ended in dismissal: in 1840 from a position as a private tutor in the Lake District, in 1842 from a job as a clerk on the Leeds-Manchester Railway, and finally and most devastatingly from his post as a tutor at Thorp Green, York, home of the Robinson family, also his sister Anne's employers. Branwell's apparent love-affair with Lydia Robinson, the lady of the house, had resulted not only in losing him his job, but also in plunging him into the depths of despair. Since his inauspicious return to Haworth in the summer of 1845, he had squandered his time on self-pity, with the occasional fruitless effort to revive his literary ambitions. More and more he depended on the solace of drink and drugs, freely obtainable over the apothecary's counter in the village. Branwell's surviving letters from his last years to his friend Leyland, the sculptor, with their frantically scrawled handwriting and gruesome self-portrait cartoons, visibly chart the deterioration of the man. Charlotte, meanwhile, increasingly found less and less sympathy for her wretched brother, and in her letters to Ellen she comments tersely on his miserable condition. For example, writing to her friend on 1 March 1847: 'Branwell has been conducting himself very badly lately – I expect from the extravagance of his behaviour and from mysterious things he drops (for he will never speak out plainly) that we shall be hearing news of fresh debts contracted by him soon – '. In fact, from the beginning of 1848, tuberculosis was taking a hold on Branwell's abused physical frame and he was not to see the year out. His last surviving letter is a scribbled note containing a pathetic plea to his village friend, John Brown, urging him to bring a measure of gin to the door of Haworth Parsonage.

Branwell Brontë died on Sunday 24 September 1848, aged thirty-one. The whole family was at his death-bed. Charlotte was deeply affected by the wastefulness of her brother's death, and she had great difficulty in coming to terms with the way she had seen her talented brother dissipate his gifts. Just a week later, writing to W. S. Williams, who had become a confidante of sorts, Charlotte actually described her brother's death as 'a mercy rather than a chastisement'. His loss, she wrote, did not give her a sense of bereavement, rather 'There is such a bitterness of pity for his life and death – such a yearning for the emptiness of his whole existence as I cannot describe – I trust time will allay these feelings.' Above all, it was the awful obscurity of his death with which she could not come to terms.

'A Parody', *Branwell Brontë's last surviving drawing, a sketch in pen and ink made just two months before his death. 'Jack Shaw the Guardsman' and 'Jack Painter of Norfolk' were boxers, referring to the metaphor used in the drawing to represent Branwell's refusal to face the challenge of death.*

The Brontë Society

He Died after 20 minutes struggle on Sunday Morning 24th Septr. He was perfectly conscious till the last agony came on – His mind had undergone the peculiar change which frequently precedes death, two days previously – The calm of better feelings filled it – a return of natural affection marked his last moments – he is in God's hands now – and the all – powerful – is likewise the all – merciful – a deep conviction that he rests at last – rests well after his brief, erring, suffering, feverish life fills and quiets my mind now . The final separation – the spectacle of his pale corpse gave more acute, bitter pain than I could have imagined – Till the last hour comes we never know how much we can forgive, pity, regret a near relation – All his vices were and are nothing now – we remember only his woes.

After Branwell's death, Charlotte wrote this letter to Ellen Nussey on 9 October 1848. In it she expresses 'a deep conviction that he rests at last – rests well after his brief, erring, suffering, feverish life'.

The British Library Ashley MS 2452, f.2v

Branwell's death left Charlotte ill, depressed and unable to write. The autumn brought cold east winds that cut down from the moors and chilled the grieving Brontës, bringing coughs and colds which, Charlotte noted fearfully, Emily and Anne seemed unable to shake off. Consumption was rife in the unhealthy village, and Charlotte was weighed down by a sense of foreboding for the well-being of her much-loved sisters. At the end of October Charlotte penned her fears to Ellen: 'Emily's cold and cough are very obstinate; I fear she has a pain in her chest – and I sometimes catch a shortness in her breathing ... she looks very, very thin and pale ... Nor can I shut my eyes to the fact of Anne's great delicacy of constitution.'

Emily Brontë was in fact very seriously ill, and what was worse for all around her, she characteristically refused to acknowledge her own frailty. As Emily continued to drag herself out of bed each day and struggle to do her household chores, Charlotte, almost demented with worry, sought counsel from all sides. Her friends and correspondents at Smith, Elder and Co. were tenderly solicitous of their celebrated writer's anguish. Both George Smith and W. S. Williams wrote frequently with whatever advice they could give, and parcels of new books were sent to distract the unhappy household. It was George Smith who recommended homeopathy to Charlotte, and gave her the name of the eminent Dr Epps, to whom Charlotte wrote a letter containing a disturbing account of Emily's symptoms. Lost for many years, this letter emerged within a collection of literary letters made by Lady Charnwood in the early years of the twentieth century, now in the British Library. Sadly, Dr Epps' advice was of no avail to the rapidly declining Emily, who continued to refuse any help until the very morning of the day of her death, which came on 19 December 1848. She was thirty years old.

In her *Biographical Notice* of her sisters, published in 1850, Charlotte gave an account of Emily's death which must stand as one of the most moving passages she ever wrote:

'My sister Emily first declined. The details of her illness are deep-branded in my memory, but to dwell on them, either in thought or narrative, is not in my power. Never in all her life had she lingered over any task that lay before her, and she did not linger now. She sank rapidly. She made haste to leave us. Yet, while physically she perished, mentally she grew stronger than we had yet known her. Day by day, when I saw with what a front she met suffering, I looked on her with an anguish of wonder and love. I have seen nothing like it; but, indeed, I have never seen her parallel in anything. Stronger than a man, simpler than a child, her nature stood alone.'

Emily Brontë's funeral card, with her age wrongly recorded. The distribution of these small, black-edged cards was a Victorian mourning custom.

The Brontë Society

The day after Emily's death, Charlotte wrote to W. S. Williams with the terrible news:

'Yesterday Emily Jane Brontë died in the arms of those who loved her. Thus the strange dispensation is completed – it is incomprehensible as yet to mortal intelligence. The last three months – ever since my brother's death seem to us like a long, terrible dream. We look for support to God – and thus far he mercifully enables us to maintain our self-control in the midst of affliction whose bitterness none could have calculated on'.

Emily was buried in the family vault beneath the church on 23 December, her funeral procession followed by her faithful dog, Keeper, who continued to keep a vigil at his mistress's door long after she had departed.

Poor Charlotte was given little time to grieve for Emily. As the new year of 1849 dawned, it became apparent that Anne, too, was suffering from the same dread symptoms, and a doctor from Leeds was called at once. Charlotte wrote once more to her publisher friend, this time expressing the awful fear that her losses had still not ended: 'When we lost Emily I thought we had drained the very dregs of our cup of trial but now when I hear Anne cough as Emily coughed, I tremble lest there should be exquisite bitterness yet to taste.' Anne at first seemed to rally, stoically taking the treatments doled out by her doctor, but it was only a brief respite, and she became ill again in March. The concerned friend, Ellen Nussey, suggested that Anne might be taken to the sea, for a change of air. Scarborough had always been a place much loved by Anne, who had spent summers there with the Robinson family, and she was considerably cheered by the proposal. Charlotte was not, and did all she could to oppose the idea, no doubt worried that the journey would accelerate her sister's demise, but she gave in to the opinion of the doctor who thought it could do the patient some good. So on 24 May Anne, Charlotte and Ellen set out for Scarborough, shopping en route in York, and visiting the Minster. Anne managed to enjoy a few of the pleasures of the resort, with

Immediately he heard of Emily's death, W. S. Williams of Smith. Elder and Co. wrote to Charlotte to express his great sympathy for her loss. Charlotte's reply, written on Christmas Day, begins: 'I will write to you more at length when my heart can find a little rest – now I can only thank you very briefly for your letter which seemed to me eloquent in its sincerity'.'

The British Library Ashley MS 2452, f.13

a visit to the spa baths and a ride in a donkey cart on the beach. For the last time the youngest Brontë viewed the glorious, sweeping North Sea bay, where in happier times she had walked, sketched and collected glistening red carnelian stones from the beach. On 28 May 1849, Anne Brontë died, aged twenty-nine years.

The decision was taken to bury Anne at Scarborough, no doubt to spare suffering all round, particularly for Patrick Brontë. So Anne was laid to rest, far apart from her mother, aunt, brother and sisters, in the churchyard of St Mary's, high above the town.

Writing a few days later to W. S. Williams, Charlotte reflected on Anne's death, and crystallised the difference in attitude she held to each of her losses, and her own survival: 'She died without severe struggle – resigned – trusting in God – thankful for release from a suffering life – deeply assured that a better existence lay before her... Anne, from her childhood seemed preparing for an early death – Emily's spirit seemed strong enough to bear her fulness of years – They are both gone – and so is poor Branwell – and Papa has now me only – the weakest – puniest – least promising of his six children – Consumption has taken the whole five.' In another letter to him, written a little later, Charlotte summed up the grief and despair of the last nine months of her life in the words of one stunned by tragedy: 'A year ago – had a prophet warned me how I should stand in June 1849 – how stripped and bereaved ... I should have thought – this can never be endured...'

Anne Brontë's grave in the churchyard of St Mary's at Scarborough, North Yorkshire. Anne was the only member of the Brontë family not to be buried at Haworth. The inscription on her gravestone gives her age incorrectly, 'She died Aged 28, May 28th 1849'.

The Brontë Society

Shirley

'If my master has given me ten talents, my duty is to trade with them, and make ten talents more. Not in the dust of household drawers shall the coin be interred. I will not deposit it in a broken-spouted tea-pot, and shut it up in a china closet among the tea things. I will not commit it to your work-table to be smothered in piles of woollen hose. I will not prison it in the linen press to find shrouds among the sheets: and least of all...will I hide it in a tureen of cold potatoes, to be ranged with bread, butter, pastry and ham on the shelves of the larder.'

Shirley, *Chapter 23*

The first page of Volume 2 of the manuscript of Shirley. *Shirley Keeldar and Caroline Helstone are the main female characters in the book, and a close friendship develops between them. Charlotte modelled the strong-willed, spirited character of Shirley on her sister Emily. As this chapter opens, Shirley is admonished for her unfeminine 'Yorkshire habit' of whistling.*

The British Library Add. MS 43478, f.1

charlotte Brontë's third novel, *Shirley*, published in 1849, was written in the most difficult of circumstances, as she suffered the intense loneliness of the loss of her brother and sisters. But as she worked she discovered the curative effects of writing, and gradually became absorbed in her tale. It is set against the richly detailed background of the West Riding of Yorkshire at the time of the Luddite riots, firmly located in the years 1811–1812. The narrative unfolds within the industrial heartland of Batley, Birstall and Dewsbury, which Charlotte knew well from her school-days at Roe Head, Mirfield, and her lifelong friendships with the Nusseys and the Taylors who lived there. At the time of her novel, the wool industry was ailing because of the lack of exports during the Napoleonic Wars and there were uprisings amongst the mass of the unemployed who were suffering great poverty. Despite the

Volume 3 of Shirley
*begins with a chapter
entitled 'The Valley of
the Shadow of Death'.
Written in the wake of
the loss of her brother
and sisters, Charlotte
creates a poignant
opening paragraph: 'The
future sometimes seems
to sob a low warning of
the events it is bringing
us, like some gathering
though yet remote
storm, which, in tones of
the wind, in flushings of
the firmament, in clouds
strangely torn,
announces a blast strong
to strew the sea with
wrecks'.*

The British Library
Add. MS 43479, f.1

politically dangerous situation, mill-owner Robert Gerard Moore, half-Belgian, half-English, is determined to introduce the new labour-saving machinery into his mill. The opposition of the workers leads to an attempt to destroy his mill, and eventually to take his life. Moore needs financial help, so he proposes marriage to the wealthy Shirley Keeldar, although he is actually in love with the more gentle and reserved Caroline Helstone, the rector's daughter, whose feelings for Moore are the same. Robert is rejected by Shirley, and when his business recovers as the war ends, he marries Caroline, his true love. Shirley, meanwhile, is drawn to Louis, Robert's brother, who is a tutor in her family. Both strong characters, their love conquers their social imbalance. Although the novel contains some strong themes, such as woman's right to work, the tale ends conventionally with the heroines finding husbands.

Charlotte was anxious about the public response to her book after the great success of *Jane Eyre*, especially as she had vowed that it would be 'as unromantic as Monday morning'. She saw her female characters to be as oppressed by their lot in society as were the mill-workers. Just as the machine-wreckers acted against the political, social and economic forces which kept them in their place, so did the women struggle against a system which perpetuated an ideal of femininity that took no account of a woman's intellectual abilities, her economic demands, or her need to expend her energies and properly express her emotions.

Charlotte sent the first volume of *Shirley* to Smith, Elder and Co., asking for an honest opinion, which she got. Her publishers were critical. The male characters were not good, they said, and they did not like the opening with the curates' supper. Charlotte, who had more experience of the behaviour of curates than most people, heatedly claimed that it was true to life. They were indeed. The models were instantly recognisable to their parishioners. In fact Charlotte drew very closely from her family, friends and acquaintances to create the numerous characters deployed

Stained-glass windows in the Red House, Gomersal, formerly the home of the Taylor family, now a museum. The Red House was recreated as Briarmains, the Yorke family home, and Charlotte describes the windows in Chapter 9: 'Those windows would be seen by daylight to be of brilliantly-stained glass – purple and amber the predominant hues, glittering round a gravely tinted medallion in the centre of each, representing the suave head of William Shakespeare, and the serene one of John Milton.'

The Brontë Society

Charlotte based many of the locations in Shirley *on her familiar knowledge of the industrialised Birstall and Batley areas of West Yorkshire. Oakwell Hall, a seventeenth-century yeoman's house, now a museum, was the inspiration for Shirley Keeldar's home, Fieldhead.*

The Brontë Society

across the canvas of *Shirley*. The eponymous heroine was to a large extent based on her sister Emily, with her remarkable strength of spirit, and it has been suggested that Caroline Helstone was to some degree a self-portrait. Certainly, the Yorke family were based on that of Mary Taylor, whose cultured, radical household is recreated, right down to the pictures decorating the rooms - 'A series of Italian views decked the walls'.

Charlotte returned to writing *Shirley* to occupy herself after her sisters' deaths. It was two-thirds written when Branwell died, and the Third Volume opens with the title 'The Valley of the Shadow' reflecting her depressed state. She wrote convincingly of the severe illness of Caroline Helstone: 'With all this care, it seemed strange that the sick girl did not get well; yet such was the case: she wasted like any snow-wreath in thaw; she faded like any flower in drought.'

It seems extraordinary that Charlotte did not realise that the models for her characters would be identified, indeed, that many would find great sport in doing so. In *Jane Eyre* she had drawn on people and places from long ago in her own past, but in *Shirley* she drew for her characters on persons still living, the inhabitants of small, closely-knit communities where everyone knew everyone else. She was surprised by the interest in her sources, but stoutly defended her fiction. What was more troubling to her was that the publication of *Shirley* renewed the public interest in the sex and identity of Currer Bell with a vengeance. 'There is woman stamped on every page' announced *The Atlas*. One review in particular cut Charlotte to the quick, written by her erstwhile correspondent George Henry Lewes, which judged *Shirley* entirely on the basis that it had been written by a woman, and its writer boasted that he knew the secret identity of Currer Bell. He exhorted her in print to 'learn also to sacrifice a little of her Yorkshire roughness to the demands of good taste'. He even referred to her spinster status by criticising the unrealistic way in which Mrs Pryor, discovered late in the novel to be the mother of Caroline Helstone, had given away her baby: 'Currer Bell! if under your heart had ever stirred a child, if to your bosom a babe had ever been pressed... never could you have imagined such a falsehood as that!'.

The manuscript of *Shirley*, along with *Jane Eyre* and *Villette*, formed part of Mrs Elizabeth Smith's bequest to the British Museum in 1914. It is considerably more reworked than the *Jane Eyre* manuscript. For example, the Sykes family were originally given the name of Roakes. Place-names, too, are in some instances altered. The hand-writing is not so firm and even as in the *Jane Eyre* manuscript, an indication of Charlotte's troubled state of mind as she wrote.

E M Wimperis' illustration of Hollow's Mill for the Smith, Elder and Co. 1872 edition of Shirley. *The model for Hollow's Mill, owned by Robert Moore, was Hunsworth Mill at Batley. When writing about the attack on the mill by the Luddites, Charlotte drew on the account of the actual attack on Rawfolds' Mill near Hartshead, where her father had been curate before she was born.*

The Brontë Society

≈ *Villette*

*'Forget him? Ah! they took a sage plan to make me forget him - the wiseheads!
They showed me how good he was; they made of my dear little man a stainless
little hero. And then they had prated about his manner of loving. What means
had I, before this day, of being certain whether he could love at all or not? I
had known him jealous, suspicious; I had seen about him tendernesses,
fitfulnesses - a softness which came like a warm air, and a ruth which passed
like early dew, dried in the heat of his irritabilities: this was all I had seen.'*

Villette, Chapter 35

Villette was published in 1853 and its title is a pseudonym for the capital
city of Belgium. Of all Charlotte's novels, it is the most consistently
autobiographical. It tells the life story of Lucy Snowe, another typical Brontë heroine
in that she lacks beauty and money. Lucy takes up a post as a teacher in a girls' school
in Brussels in order to support herself. She has a strong character, is stoical in the face
of adversity and works hard, enough to win the respect of the capable yet
unscrupulous headmistress, Madame Beck, an unattractive character whom
Charlotte based on her own ex employer Madame Claire Zoë Heger. Lucy is
strongly attracted to the handsome young English doctor, John Graham Bretton, but
represses her feelings and observes his infatuation with one of her students, the
flirtatious Ginevra Fanshawe, from a distance. She is relieved when he finds a happier
love with Paulina Home, his friend from childhood. Charlotte clearly based the
sympathetic Dr John on George Smith, her publisher.

The central them of the novel, however, is Lucy's gradual fascination with the
sharp-tongued, despotic but kindly little professor, Monsieur Paul Emanuel, a figure
based on Constantin Heger, the object of Charlotte's unrequited love when she too
was a teacher in Brussels. Monsieur Paul's feelings towards Lucy are transformed
from bitterness and tyranny into respect and warm affection. Generously, he sets
Lucy up as the mistress of her own school in Brussels when he has to leave for the

Chap. 26th

The little Countess

Cheerful as my god-mother naturally was, and entertaining
for our sakes, she made a point of being, there was no true
enjoyment that evening at La Terrasse till — through the
wild howl of the winter-night — were heard the signal sounds of
arrival. How often, while women and girls sit warm at snug
firesides, their hearts and imaginations are doomed to divorce
from the comfort surrounding their persons, forced out by
night to wander through dark ways, to dare stress of weather,
to contend with the snow-blast, to wait at lonely gates and
stiles in wildest storms, watching and listening to see and
hear the father, the son, the husband coming home.

Father and Son came at last to the chateau, for the
Count de Bassompierre that night accompanied Dr. Bretton.
I know not which of our trio heard the horses first; the as-
perity, the violence of the weather warranted our running down
into the hall to meet and greet the two riders as they came
in; but they warned us to keep our distance; both
were white — two mountains of snow; and indeed

West Indies on business. Whether he shall live to return and marry her is left for the reader to decide in a tantalisingly ambiguous last few paragraphs.

When she wrote *Villette* Charlotte was very conscious of the fact that it would be read by a public which now knew of her real identity. Her publishers were aware of this too, and remonstrated with her to soften the character of Lucy Snowe, whom they regarded as too cold and mysterious. Charlotte was torn between calling her heroine 'Lucy Snowe' or 'Lucy Frost' – 'A cold name she must have' – and resolutely defended her decision to show up Lucy's faults as well as her virtues. She laboured hard at *Villette*, working with much revision, as the manuscript itself reveals, with its heavily inked deletions and the sometimes mutilated pages.

The reviews, when they came, were favourable yet subdued. Some critics noted the rather cynical and bitter spirit of the book. Once more, as with Lewes' personally cruel review of *Shirley*, Charlotte was dismayed to find herself betrayed by one whom she had believed to be her friend. Harriet Martineau, writing in the *Daily News*, used her intimate knowledge of the author to inform her remarks on *Villette*, describing it as: '... almost intolerably painful... so incessant is the writer's tendency to describe the need of being loved, that the heroine, who tells her own story, leaves the reader at last under the uncomfortable impression of her having entertained a double love, or allowed one to supersede another without notification of the transition.'

Mercifully, Charlotte never read Thackeray's thoughts on *Villette*, which he wrote privately, in a letter to Lucy Baxter 11 March 1853:

> '*it amuses me to read the author's naive confession of being in love with 2 men at the same time; and her readiness to fall in love at any time. The poor little woman of genius! the fiery little eager brave tremulous homely-faced creature! I can read a great deal of her life as I fancy in her book, and see that rather than have fame, rather than any earthly good or maybe heavenly one that she wants some Tomkins or another to love her and be in love with. But you see she is a little bit of a creature without a penny worth of good looks, thirty years old I should think, buried in the country, and eating up her own heart there, and no Tomkins will come. You girls with pretty faces and red boots (and what*

Opposite page:

On this page of Volume II of the manuscript of Villette, *Charlotte opens her chapter 'The Little Countess' with a tongue-in-cheek description of how men must work and women must wait: 'How often, while women and girls sit warm at snug firesides, their hearts and imaginations are doomed to divorce from the comfort surrounding their persons, forced out by night to wander through dark ways, to dare stress of weather, to contend with the snow-blast, to wait at lonely gates and stiles in wildest storms, watching and listening to see and hear the father, the son, the husband coming home.'*

The British Library
Add. MS 43481, f.199

Charlotte created a fascinatingly complex character in the form of M. Paul Emanuel, described here on the manuscript page opening the chapter 'M. Paul': 'I used to think, as I sat looking at M. Paul, while he was knitting his brow or protruding his lip over some exercise of mine, which had not as many faults as he wished (for he liked me to commit faults: a knot of blunders was sweet to him as a cluster of nuts), that he had points of resemblance to Napoleon Bonaparte. I think so still.'

The British Library Add. MS 43482, f.45

not) will get dozens of young fellows fluttering about you - whereas here is one a genius, a noble heart longing to mate itself and destined to whither away into old maidenhood with no chance to fill the burning desire.'

Charlotte would have burned with anger at Thackeray's description of her as a desperate old spinster, and would have no doubt rounded on him for taking her fiction to be her own life story in every detail. As for all writers, her own experiences formed the basis of her writing, but we must never forget that what she wrote was fiction. Charlotte had planned an unhappy ending for the book, with Monsieur Paul drowned at sea, but the story goes that her father, asked for his advice, pleaded for something happier, hence the ambiguity of the ending: 'Here pause: pause at once.

There is enough said. Trouble no quiet, kind heart; leave sunny imaginations hope. Let it be theirs to conceive the delight of joy born again fresh out of great terror, the rapture of rescue from peril, the wondrous reprieve from dread, the fruition of return. Let them picture union and a happy succeeding life.'

For some readers, *Villette* is regarded as Charlotte's greatest novel. It contains some remarkable passages of writing: the gothic atmosphere of buildings and gardens, Lucy wrestling with Roman Catholicism, her hallucinatory tour of Brussels, an astringent comment on the representation of women in paintings. In many ways a tour de force, *Villette* was to be Charlotte Brontë's last novel.

E. M. Wimperis' illustration of the Pensionnat de Demoiselles in the Smith, Elder and Co. 1872 edition of the novel.

The Brontë Society

*Sainte Gudule,
Brussels. Charlotte
recalled Sainte Gudule
when she wrote of Lucy
Snowe's desperate visit
to the church of Sainte
Jean Baptiste, in search
of comfort in her
nervous illness during
the long vacation.*

The Brontë Society

*Portrait of Mlle Rachel,
Actress, 1840,
William Etty.
When in London in
1851, Charlotte saw the
French actress perform,
and recreated her as
'Vashti' in* Villette: *'I
had heard this woman
termed 'plain', and I
expected bony harshness
and grimness -
something large,
angular, sallow. What I
saw was the shadow of a
royal Vashti: a queen ,
fair as the day once,
turned pale now like
twilight, and wasted
like wax in flame.'*

York City Art Gallery

⪘ *Fame*

Although Charlotte Brontë craved literary success, she did not yearn for fame, and when its rewards and its trials came her way, she found it difficult to cope with both. The sisters had made a pact of anonymity, one that Charlotte rather grudgingly accepted during her sisters' lifetimes, and then found hard to give up when she found herself on her own. She even kept the publication of *Jane Eyre* from her own father, and did not present him with her achievement until she could press a copy of the book into his hands. The story survives of how Patrick, having read the novel, announced to the family: 'Children, Charlotte has been writing a book - and I think it is a better one than I expected.' In the wider world, the fame of *Jane Eyre* moved at a brisker pace. In February 1848 it was adapted for the stage by John Courtney, it was translated into French and the American edition of the novel came not far behind. Revenue from the sales increased, and Charlotte received a steady income from her writing. A bonus of her fame was the upsurge in her correspondence, not only with W S Williams at Smith, Elder and Co., but also with other notable writers and critics, including George Henry Lewes, W M Thackeray and Julia Kavanagh. Parcels arrived at the parsonage, too, containing gifts of books from various publishers, eager for her opinion, so Charlotte was able to read the newest books and feel in touch with the literary world. Engrossed with her more celebrated correspondents, there were times when Charlotte wrote less to her loyal friend Ellen, and when she did she said nothing of her new-found fame. Ellen, excluded from her friend's secret, inevitably began to suspect her of the authorship of *Jane Eyre*, but until the sad demise of Emily and Anne, she had to remain outside of Charlotte's confidence.

Currer, Ellis and Acton Bell could not protect their identities forever, and the secret began to filter out before the deaths of Charlotte's two younger sisters. The publication of Anne Brontë's second novel, *The Tenant of Wildfell Hall*, acted as a catalyst, and events leading up to it persuaded Charlotte and Anne to take the drastic step of making a trip to London in July of 1848. It was provoked by the actions of the ever unscrupulous publisher, Thomas Cautley Newby. Despite Charlotte's best efforts to persuade Emily and Anne to abandon Newby after the difficulties surrounding the publication of *Wuthering Heights* and *Agnes Grey* in December 1847, and sign up with Smith, Elder and Co., they had resisted. Anne entrusted her second novel to Newby, and there has ever since been speculation, based on the existence of a letter from Newby found in Emily's writing-desk, that she too had begun another novel. Newby sent an extract from *The Tenant of Wildfell Hall* to the United States, claiming that it was by Currer Bell. George Smith heard of this and understandably felt betrayed. There was nothing for it but for Charlotte and Anne to go to London to prove that they were two separate persons.

The two Brontë sisters travelled by train through the night, found their way to the Chapter Coffee House on Paternoster Row – where Charlotte had stayed six years earlier en route to Brussels – washed, breakfasted and then, full of trepidation,

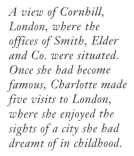

A view of Cornhill, London, where the offices of Smith, Elder and Co. were situated. Once she had become famous, Charlotte made five visits to London, where she enjoyed the sights of a city she had dreamt of in childhood.

The Brontë Society.

set off immediately for the offices of Smith, Elder and Co. at 65 Cornhill. George Smith, in his Memoir of 1902, later described his first sight of Charlotte Brontë:

'I must confess that my first impression of Charlotte Brontë's personal appearance was that it was interesting rather than attractive. She was very small, and had a quaintly old-fashioned look. Her head seemed too large for her body. She had fine eyes, but her face was marred by the shape of the mouth and by the complexion. There was but little feminine charm about her; and of this fact she herself was uneasily and perpetually conscious. It may seem strange that the possession of genius did not lift her above the weakness of an excessive anxiety about her personal appearance. But I believe she would have given all her genius and her fame to have been beautiful.'

Although his astute account was written many years later, with the benefit of knowledge of Charlotte's character and behaviour, much of this must have been conveyed by the first sight of the tiny, pale-faced woman in her provincial clothes, peering up at him nervously through her spectacles. Charlotte's first impression of Smith was of a 'young, tall, gentlemanly man', and of her long-time correspondent W. S. Williams as 'a pale, mild, stooping man of fifty'.

George Smith was eager to fling his two anxious visitors into the maelstrom of London literary society, conscious of what a stir it would create for the world to discover that the coarse Bell brothers were in fact these nervous little women from Yorkshire, but Charlotte halted him at once with her grave determination that the truth of the Brontës' identity would remain with the publishers.

Thwarted in his attempt to show them off, Smith nevertheless could not be dissuaded from sweeping off his charges on a whirl of social activity. In a daze, Charlotte and Anne found themselves whisked off to the opera, and on visits to the National Gallery and the Royal Academy of Arts. Even as they toured London, the reviews of *The Tenant of Wildfell Hall* began to appear, with the *Spectator* declaiming 'There is a coarseness of tone throughout the writing of all these Bells, that puts an offensive subject in its worst point of view, and which generally contrives to dash indifferent things'. Inevitably, the publication of Anne's second novel caused the critics to return to the Bells' works, this time very much on the attack and warning young ladies not to read them.

Whilst they were together, the Brontës kept their identities secret, and supported each other in the face of adverse criticism. Without her sisters' support, the secret lost its appeal for Charlotte, but it was nonetheless difficult to reveal

George Smith was Charlotte's publisher and friend at the firm of Smith, Elder and Co. Smith took over the family firm in 1846 and transformed it into a highly respected, thriving concern. He was only twenty-three years old when he accepted Jane Eyre, *and Charlotte relied on his shrewd literary judgement and his moral support throughout her career.*

The Brontë Society.

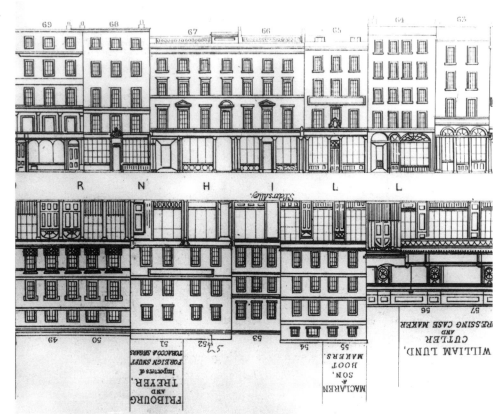

herself as the author Currer Bell. As she emerged slowly from the immediate grief at the deaths in the family, she realised that her intellect and her writing were all she had left, and returned to writing *Shirley*, two-thirds completed when Branwell died. Although she was preoccupied with her own and her father's health the manuscript was ready in August 1849. After the publication of *Shirley*, Charlotte had to reconcile herself to the fact that sooner or later she would be unmasked in the eyes of her readership. There was an ever-increasing interest in the true identity of Currer Bell, and furthermore the critics had by then decided that *Shirley* must have been written by a woman. But friendships were forged out of the public revelation. Charlotte was heartened by a warm letter from Elizabeth Gaskell, to which she replied with honesty and respect, and so began a major literary relationship. In November 1849, as her strength returned, Charlotte decided that she would go to London to stay with George Smith and his mother in Westbourne Place, Paddington, and revive her ambition.

The advantage of fame that Charlotte most appreciated was the opportunity it provided to see the sights of the great city which she had loved in her imagination for so long. Most gratifyingly, at last she could view the work of the artists she had admired since childhood. Her sightseeing on this visit took her to see private collections of work by the great British landscape artist, J. M. W. Turner. She declared that she preferred his water-colours to the later oils, which baffled her with their impressionistic effects. She went to the theatre, and saw William Charles Macready in *Macbeth* and *Othello*, and she visited the new Houses of Parliament designed by Charles Barry. The night of 4 December 1849 confronted her with an awful social ordeal, a dinner party at the Smiths which included Thackeray among the guests. So anxious was she at the prospect, she could not eat all day. In a letter to her father she described her view of her hero: 'yesterday I saw Mr Thackeray... He is a very tall man – above six feet high, with a peculiar face – not handsome – very ugly indeed – generally somewhat satirical and stern in expression, but capable also of a kind look.' On Sunday 9 December she visited Harriet Martineau, the novelist and author of essays on political economy. Charlotte made a great impression on Martineau with her powerful intellect and her physical appearance, 'the smallest creature I had ever seen (except at a fair) and her eyes blazed.'

Currer Bell's secret identity was out, and the news spread rapidly through London circles. It was with some relief that Charlotte fled back to Yorkshire for the first anniversary of Emily's death, turning once more to Ellen Nussey for comfort, who was herself beginning to enjoy basking in her friend's reflected glory. Home would soon cease to be a refuge any longer, as word got out locally. Patrick Brontë proudly began to tell his friends of his daughter's success, and by February 1850, everybody in Haworth knew that the rector's daughter was in fact the famous authoress, Currer Bell. Copies of *Jane Eyre* and *Shirley* were bought for the Mechanics' Institute in Keighley and were much fought over, and the first curiosity hunters found their way to the door of Haworth Parsonage. To confirm it all, the *Bradford Observer* reported 'It is understood that the only daughter of the Rev. P Brontë, incumbent of Haworth, is the authoress of *Jane Eyre* and *Shirley*, two of the most popular novels of the day, which have appeared under the name of 'Currer Bell'.'

Charlotte suffered a sense of anti-climax after the publication of *Shirley* and her visit to London, made worse by the absence of letters from her friends at Smith, Elder and Co. She became depressed. Her low state of mind was not helped by the ever-pressing and uninvited attentions of people who wished to share the company of such a famous person. Notable among these was Sir James Kay-Shuttleworth, who harried Charlotte with persistent invitations to visit Gawthorpe Hall, his home near Burnley in Lancashire. Kay-Shuttleworth was a medical doctor who had been Secretary to the Committee of the Council on Education, retired because of ill health and had been made a baronet. Charlotte found her admirer impossibly over-bearing, but she was entranced with the beauty of Gawthorpe itself, with its plasterwork and oak-panelling and the view of Pendle Hill, a promontory which had featured in the landscape of the Angrian stories of her youth.

In the spring of 1850, Charlotte found herself suffering a renewed sadness for the loss of her sisters. Writing did not come easily and she turned to her letters and parcels of books for comfort. She read the novels of Jane Austen, and pronounced on them with her usual acuity:

> '*She does her business of delineating the surface of the lives of genteel English people curiously well; there is a Chinese fidelity, a miniature delicacy in the painting: she ruffles her reader by nothing vehement, disturbs him by nothing profound: the Passions are perfectly unknown to her*'.

Charlotte found it harder to deal with the irritating demands of aspirant writers, whose letters seeking her advice arrived with increasing frequency. It was time for another visit to London to stay with the Smiths, one that brought Charlotte many long-awaited pleasures. As well as visits to the opera, to the Royal Academy exhibition, where she admired Landseer's portrait of Wellington and John Martin's *The Last Man*, there were expeditions to the Zoological gardens, the Ladies Gallery of the House of Commons, to the Chapel Royal, where she glimpsed The Iron Duke himself, and to the Friends Meeting House, where Charlotte revealed a lifetime's insensitivity to other people's forms of worship by finding amusement in the Quaker prayer meeting. She met Thackeray once more, going to his house for dinner on 2

June, when a bevy of women writers had been invited to meet her. Thackeray's daughter Anne described Charlotte's appearance at this gathering:

'a tiny, delicate, serious, little lady, pale, with fair straight hair, and steady eyes. She may be a little over thirty; she is dressed in a little barege dress with a pattern of faint green moss. She enters in mittens, in silence, in seriousness.'

It was also noticed that Charlotte wore a rather obvious hairpiece, the result of some misplaced fashion advice from Ellen Nussey. The ladies in the drawing-room were disappointed by her lack of brilliant conversation. Indeed, the famous writer was reported to have spent the evening in a corner, murmuring quietly to the governess.

Fame brought many ordeals, one of which was having her portrait drawn in chalks by George Richmond, who portrayed all eminent Victorians. For one so conscious of her own shortcomings it must have taken immense courage to agree to the sitting. She was very nervous, and when the artist mistook her hairpiece for a cap and asked her to remove it, Charlotte was reduced to tears of embarrassment. The resulting portrait no doubt flattered her actual appearance, but Richmond caught the intelligent beauty of her eyes, so often commented upon, and his image of Charlotte Brontë has become one of the best known portraits of a writer ever produced. It was commissioned by George Smith as a gift for Patrick Brontë, and it hung in pride of place above the fire-place of the dining-room of the parsonage. Unable to forget the experience of sitting for Richmond, Charlotte turned down other requests to paint her, including one from the Pre-Raphaelite artist John Everett Millais.

This visit to London lasted a whole month, and it was followed by a visit to Edinburgh with George Smith and his sister Eliza, which became a Walter Scott pilgrimage, with a visit to Abbotsford, to the ruined abbey at Melrose, to Scott's monument and Arthur's Seat. Charlotte had spent five weeks away from Haworth, and her return home was soon followed by the arrival of George Smith's gifts of Richmond's portrait for the Reverend Brontë, and for Charlotte an engraved portrait of the Duke of Wellington. George Smith's attentions led Ellen Nussey to believe that he was her friend's suitor, a possibility which Charlotte hotly denied.

Friendships came out of Charlotte's fame, as well as unwanted attentions. On 19 August 1850 she went to Windermere to stay with the Kay-Shuttleworths and there met Mrs Gaskell, who noted Charlotte's good features as well as her physical inadequacies:

> '*thin and more than half a head shorter than I, soft brown hair, not so dark as mine; eyes (very good and expressive looking straight & open at you) of the same colour, a reddish face, large mouth & many teeth gone; altogether plain; the forehead square, broad, and rather overhanging. She has a very sweet voice, rather hesitates in choosing her expressions, but when chosen they seem without an effort, admirable and just befitting the occasion.*'

Charlotte warmed to Elizabeth Gaskell, 'I was truly glad of her companionship. She is a woman of the most genuine talent – of cheerful, pleasing and cordial manners and – I believe – of a kind and good heart.' She saw much of the Lake District scenery on this visit, all the places she had once only known from poor engravings in books of views. Charlotte longed to go off and roam free, but she had to be driven around and feted by the Kay-Shuttleworths, showing off their celebrated authoress.

As the only surviving Brontë, Charlotte felt the need to become both custodian and censor of her sisters' literary output, and their life histories. Later commentators have criticised her for this, seeing her as a manipulative force. However, it was inevitable that she should be consulted when Smith, Elder published cheap one-volume versions of *Wuthering Heights* and *Agnes Grey*. Charlotte leapt at the chance to write a preface, and added some of her sisters' poems, which she edited in what could be described as a high-handed fashion. The root of her motives must be found in her unwavering sense of loss and a determination to protect her sisters' posthumous reputations. Charlotte's loneliness got the better of her for long periods, when she was unable to write. It was a long time before she could bring herself to start work on her next novel, *Villette*.

Visits to London continued to provide her with a means of escape from the oppressive atmosphere of the parsonage. On 29 May 1851 she embarked on her

busiest and most public visit, when she attended a lecture by Thackeray on the English Humorists of the eighteenth century. Thackeray embarrassed her by introducing her to his mother as 'Jane Eyre', for which she took him to task. Charlotte also made a number of visits to the greatest tourist attraction of the day, the Great Exhibition, where she was struck by the great cornucopia of objects on display. Also memorable was an outing to the French Theatre, where she saw 'Rachel', the most famous actress of her day, whom she recreated as Vashti in *Villette*. Charlotte resisted all attempts to show her off to society ladies and found much more pleasure in her expeditions with George Smith, one of them a visit to a fashionable phrenologist in the Strand, who read character from the bumps and indentations in the cranium. They called themselves 'Mr and Miss Fraser' for the purposes of the visit. Charlotte's reading described a nervous temperament, sensitive nature, a tendency to gloom, high ideals and poetical sentiments. Her large forehead was interpreted as a sign of deep thoughtfulness, great intellect and a fine command of language.

Charlotte Brontë's fame brought her much attention, some of which she desired, some she merely tolerated. Her friendship with George Smith enlivened her, and it has frequently been speculated that she had expectations of him. If she did, her hopes were abruptly dashed when she learned of Smith's engagement to Elizabeth Blakeway, the daughter of a London wine merchant, in November 1853. However, throughout all of this time a shadowy figure in her life had been building up a strong affection and respect for Charlotte, a man who was to introduce her to happiness as her life drew to a close.

Opposite page:

The engraved portrait of William Makepeace Thackeray which was a gift to Charlotte from George Smith. The author of Vanity Fair *was one of Charlotte's literary heroes, although when she first met him she was overcome by her habitual shyness.*

The Brontë Society.

⟨⟩ *Marriage*

Arthur Bell Nicholls first came into Charlotte's life in 1845 when he was appointed curate at Haworth, but he remains a shadowy background figure in the Brontë story, almost until the end. Born in January 1819 near Belfast in Northern Ireland, he was the son of a poor farmer, with a background not dissimilar to Patrick Brontë's own, as he was sponsored by his uncle to attend Trinity College, Dublin and then went on to be ordained into the church. He was twenty-six years old when he came to Haworth, a tall, solid man with a square face, dark hair and side-whiskers. Hard-working, dutiful and placid on the surface, he proved to have deeply hidden emotions which made him capable of great outbursts of feeling, a nature which eventually made him more interesting to Charlotte. Her first impression of him was as a respectable young man who read well in Church. It is quite likely that his middle name was the inspiration for the Brontës' pseudonym 'Bell'. He apparently had a sense of humour, too, given the account of how he laughed heartily at Charlotte's portrayal of the curates in *Shirley*, not affronted to detect something of himself in them.

From his arrival in Haworth, unsurprisingly, Nicholls was a frequent visitor to the parsonage, which led to gossip in the ever-watchful village that the curate was courting the rector's eldest daughter. This local speculation probably created Charlotte's hostile feelings towards him in the early part of his time in Haworth. Any mentions she made of him in her letters to Ellen were invariably withering in tone. And yet he was a tower of strength for her father, taking on the lion's share of the work of the parish as Patrick Brontë's health faltered. It was Nicholls who officiated at Emily Brontë's funeral. He must have been sensitive to Charlotte's great grief at the loss of her brother and sisters, and gradually developed an affection for her. However, it was not until 13 December 1852 that Arthur Bell Nicholls asked Charlotte Brontë to marry him.

Charlotte was completely taken aback by the strength of his feelings. In a letter to Ellen she described how he stood before her 'Shaking from head to foot, looking deadly pale, speaking low, vehemently yet with difficulty – he made me for

Arthur Bell Nicholls was born in Northern Ireland and was curate at Haworth from 1845 until 1861. Nicholls and Charlotte Brontë were married in 1854, having overcome much opposition to the idea from Patrick Brontë.

The Brontë Society.

the first time feel what it costs a man to declare affection where he doubts response.' She was equally shocked by her father's outraged reaction to the news – 'Agitation and Anger disproportionate to the occasion ensued'. Nicholls' proposal knocked all of her preconceived notions of his character out of Charlotte's head, and at this point her feelings towards him gradually changed from hostility to kindliness. Dutiful daughter as ever, she turned Nicholls down, overwhelmed by Patrick's intense anger at the proposition. Patrick Brontë objected to the fact that Arthur Nicholls had not sought his permission to ask for Charlotte's hand. Also, he saw him as a fortune-hunter, believing that his motive was the considerable wealth Charlotte had amassed from her writing. Although it was not voiced, her father must also have been concerned at the almost inevitable consequences of marriage, that frail Charlotte, well into her thirties, would not survive pregnancy. He could not bear the prospect of losing his only surviving child.

Unable to bear Nicholls' anguish at her refusal, her father's wrath and his increasingly vindictive treatment of his curate, Charlotte fled to London in January 1853. On her return, she discovered that her would-be suitor had felt compelled to get away from the site of his disappointment and apply for work as a missionary in Australia. However, he eventually withdrew his application and sought another curacy instead. On 25 May 1853 Nicholls left Haworth for a new post at Kirk Smeaton, Pontefract, and this time Charlotte felt sorrow for him, becoming racked with misery and guilt at his treatment. She became ill with influenza, and to add to her woes Patrick Brontë had a stroke, which badly damaged his sight. Nonetheless, she told her father that although she had submitted to his wishes, and refused Arthur Nicholls, she felt that he had been cruel to him. Patrick Brontë had perhaps unwittingly caused Charlotte to see Nicholls in a more attractive light – as the tormented lover, a character she was drawn to in her own fiction. Charlotte's gloom over the whole affair was exacerbated by her friend Ellen Nussey's opposition to Nicholls, no doubt jealous of the place the man was to take in her dear friend's affections. Not to mention the discovery that George Smith was engaged to be married. Estranged from her father, her oldest friend and her admired publisher, she must have felt lonely indeed.

Charlotte Brontë's wedding bonnet and veil. Those who saw Charlotte on her wedding-day said that she looked like a little snowdrop in her white gown decorated with green flowers and embroidery.

The Brontë Society.

The dress which Charlotte wore on her honeymoon tour is made of finely striped lavender and silver silk, now faded to brown. The gown has a tiny waist, achieved by tight corseting. George Smith commented in his Memoir: 'she had I should say too small a waist. It shocks the respect for a fine genius to say it; but I have no doubt that tight-lacing shortened Charlotte Brontë's life.'

The Brontë Society

On 19 September 1853, Elizabeth Gaskell came to stay at Haworth. Charlotte told her the whole story of Nicholls' proposal and her father's animosity. It was unfortunate that Gaskell met Charlotte and her father at such a low point in their relationship, an accident of fate which was to colour her account of him as austere and unyielding in her *Life of Charlotte Brontë*, an image which was to influence most later accounts of the Brontës' history. The novelist was keen to intervene and do something to improve the chance of a marriage between Charlotte and her curate. She wrote to her friend Richard Monckton Milnes, a literary collector and admirer of Charlotte's work, to ask whether, with his involvement in various charitable trusts, he could obtain a pension for Arthur Nicholls and therefore make the marriage a possibility. This was achieved, much to Nicholls' bewilderment, who had no idea how such bounty had come his way.

Early in 1854, Nicholls stayed with Mr Grant, the curate of Oxenhope, and managed to meet with Charlotte several times, a fact which she at first concealed from her father. Eventually, sickened by the subterfuge, she confessed about the meetings to Patrick and demanded that she may be allowed to meet openly with Nicholls. Patrick Brontë had to give in, and in April 1854 Nicholls spent a week at the parsonage and renewed his proposal. This time he was accepted, and Patrick agreed to the marriage, as long as Nicholls lived at the parsonage and worked as his curate once more, so that he would not be separated from his daughter. At last the engagement was announced, and Charlotte wrote to tell her friends the news. She wrote in a curiously melancholy tone, admitting to subdued expectations of marriage. This was to be no Jane Eyre and Rochester love-match, and well she knew it. Cares and fears were mixed up with the cautious personal hopes of a novelist who had won fame and fortune by writing about unconquerable passion.

Preparations for the wedding were made. On 22 May 1854 Nicholls came to the parsonage to draw up the marriage settlement, a common practice in the days before the married Women's Property Act, when all the woman's property automatically became her husband's unless she made a legal agreement stating otherwise in advance of the marriage. Charlotte's wealth amounted to £1,678 9s 9d, composed of her earnings and her railways investments. The settlement she made confirmed that if she died childless before Nicholls her estate would revert to her father, not to her husband. For the rest of her life she would draw an income from her money for her 'sole and separate use'. Nicholls was therefore excluded from her wishes. The wedding plans were made very discreetly, as the couple did not wish to be a spectacle for the whole village. They were married on 29 June in the church of St Michael's and All Angels, Haworth, with the Reverend Sutcliffe Sowden officiating. Patrick decided not to attend, and Miss Wooler, Charlotte's former teacher, gave her away, with Ellen as bridesmaid. At eight o'clock in the morning the small party walked to the church for the ceremony. Charlotte wore a demure white muslin dress with green embroidery, a lace mantle and a white bonnet trimmed with lace and flowers. Those who saw her said she looked like a snowdrop. After a wedding breakfast at the parsonage, Mr and Mrs Nicholls set off to North Wales to catch the packet steamer bound for Dublin, where they spent two days sight-seeing. Then it was on to Banagher, Nicholls' home town, where Charlotte was gratified to discover that Arthur's origins were not so humble after all, with a large family home and genteel relations. Light-hearted letters survive in which Charlotte declaims the happiness she felt on her honeymoon tour of the west coast of Ireland. Even an incident in Cork, where she fell off a horse and was nearly trampled to death, did not detract from her enjoyment. Interestingly, the newly-weds did not make any visits to members of Patrick Brontë's family, who lived in Northern Ireland.

The couple were back home by 1 August, and Charlotte's letters continued to reveal her new-found happiness, despite the difficulties she sometimes found after a lifetime of single-minded independence. Indeed, she soon found great pleasure in the daily round of the married woman's life. Nicholls was always protective of his wife's privacy, and his own, and Ellen Nussey was dismayed to hear from Charlotte that her husband expected her to destroy Charlotte's letters. Although she promised to do this, clearly she did not keep her word. As the year drew to a close, and the anniversary of Branwell's and Emily's deaths approached, Charlotte was for the first time able to bear the prospect with some sense of peace. There was one other significant difference in her life. She had stopped writing. The last thing she wrote was the introductory fragment to what would have been her fifth novel, *Emma*.

Charlotte Brontë never expected to marry, nor did she imagine that she would ever enjoy the love and companionship which embraced her marriage. But it was not to last, for in January 1855 Charlotte caught a chill from which she never recovered. She was also in the early stages of pregnancy and suffering badly from nausea. By the end of the month she was confined to bed and the doctor was sent for. Constant vomiting rapidly diminished her strength, and on 17 February she made her will. In it the marriage settlement was overturned and everything was left to her beloved Arthur. No provision was made in writing for her father as she knew that Nicholls would look after him. A few weakly pencilled notes survive from the hand of one of the century's most avid correspondents as her life slipped away. Just three weeks before her thirty-ninth birthday Charlotte Brontë died, early in the morning of 31 March 1855, her unborn child dying with her.

Arthur Bell Nicholls' letter to Ellen Nussey, written 31 March 1855, the very day of Charlotte's death, telling her the sad news.

The Brontë Society.

Opposite page:

A collection of mourning items and Charlotte Brontë's funeral card.

The Brontë Society.

In Memory of

CHARLOTTE NICHOLLS,

WHO DIED MARCH XXXI, MDCCCLV,

Aged 38 Years.

⬿ *Afterlife*

Almost as soon as Charlotte had been laid to rest in the family vault beneath the Haworth church, prurient speculation about her life history began in the press. A particularly lurid article in *Sharpe's London Magazine*, brought to the attention of Patrick Brontë and Arthur Nicholls by Ellen Nussey, provoked Patrick to take up Ellen's suggestion that he ask Elizabeth Gaskell to write a biography of Charlotte that would put the record straight. The great irony in this was that, unbeknown to either of them, it was Mrs Gaskell herself who had been the source of the offensive piece in the magazine. Charlotte's husband and father were both worried by the overwhelming interest in her life history, Arthur in particular, who had always had a highly protective attitude to his wife's reputation. But already the great machine of Brontë myth was beginning to roll.

Elizabeth Gaskell fell to her task with energy, determination and great speed. She enlisted the help of all those friends and acquaintances she could find, depending especially on Ellen Nussey, who was now coming into her own as the fount of all knowledge of Charlotte's life. Nussey gave Gaskell over three hundred of Charlotte's letters, the ones that Nicholls believed to have been destroyed, hurriedly censoring them to remove the names of those whom she felt should remain anonymous. Miss Wooler, George Smith, W. S. Williams, old school-friends, ex-employees of the Brontës, village gossips – all were importuned to give up their letters and their memories. Even Mary Taylor was contacted in New Zealand, although she had disposed of all of Charlotte's letters. The overbearing Kay-Shuttleworth was swiftly on the scene, and practically ambushed Brontë and Nicholls in their own home, eager to help extract as many letters and manuscripts as he could to assist Gaskell's researches.

So, in 1857, within two years of Charlotte's death, Elizabeth Gaskell's *Life of Charlotte Brontë* was published. Any hopes that father and widower might have had of discretion and restraint in the book were dashed to pieces. Gaskell had written a colourful tale which sanctified Charlotte Brontë, making her into a martyr of terrible circumstance, thus giving birth to a whole school of Brontë biography which continues to this day. Charlotte the intellectual, the independent spirit, the manipulative sister and the cynical commentator were lost to the one image of the unhappy victim of a tragic life. From that day onwards, Brontë tourism was born, as ardent fans found their way to the village of Haworth, to be met by villagers eager to share their reminiscences and show them their Brontë relics, at a price. In the same year Charlotte's first novel *The Professor* was finally published, but was overshadowed by the much more captivating tale of her own life.

On 7 June 1861, Patrick Brontë died aged eighty-four, far outliving all of his children. Arthur Nicholls, who had loyally looked after his father-in-law in his declining years, and carried out the duties of curate with quiet conscientiousness, was callously passed over for the post of rector and returned to Banagher in Ireland. He took with him as many of Charlotte's personal and literary possessions as he could save from the hands of the souvenir hunters. The rest went into the sale of household effects held at the parsonage. Nicholls later remarried, and lived to the great age of eighty-eight, for over forty years acting as quiet custodian of many important Brontë relics. If pressed, he was willing to talk or write with pride of his celebrated first wife, but he nevertheless remained averse to publicity until the end of his days

The new rector, John Wade, possibly appointed partly for his lack of interest in the Brontës, added a new wing to the parsonage in 1872–1878. In 1879 he was instrumental in having the church demolished, except for the tower, and rebuilt. In 1880 Martha Brown, the Brontës' long-time servant died, leaving a sizeable collection of Brontë material, some of which was sold at Sotheby's, as Nicholls' material was to be, and the rest was handed down in her family. More and more collectors became interested in acquiring Brontë manuscripts and memorabilia, and in 1893 the Brontë Society was founded, one of its aims being to preserve such items with the ambition of opening a museum dedicated to the extraordinary literary family. Their first enterprise was a small Brontë Museum in a room above the Yorkshire Penny Bank at the top of Haworth Main Street, which opened in 1895. Two years later Ellen Nussey died, aged eighty, having been the victim of some of the more unscrupulous Brontë scholars who had conned her out of her Brontë letters

Charlotte Brontë

The Brontë Parsonage Museum at Haworth, West Yorkshire. Today the former home of Charlotte Brontë attracts tens of thousands of visitors a year, eager to see the place where some of the greatest novels in English literature were written.

The Brontë Society

WILLIAM HURT　CHARLOTTE GAINSBOURG　JOAN PLOWRIGHT　ANNA PAQUIN

JANE EYRE

A FILM BY FRANCO ZEFFIRELLI

Poster for the 1996 film version of Jane Eyre *by Franco Zeffirelli, starring Charlotte Gainsbourg and William Hurt. Film-makers continue to take up the challenge of transferring this classic of English literature from printed page to the screen.*

Guild Entertainment Ltd

with a false promise that they would be given to the nation. In fact, the five hundred or so letters she had so carefully preserved are now distributed widely around the world.

In the early years of the twentieth century, some of the most significant Brontë items found their rightful homes. In 1914 Elizabeth Smith's bequest of Charlotte's manuscripts was made to the British Museum. Around the same time, Branwell's youthful portraits of his sisters went into the collection of the National Portrait Gallery in London. Then in 1928, the parsonage at Haworth was bought by a local businessman, Sir James Roberts, who presented it to the Brontë Society, and it opened as the Brontë Parsonage Museum. In North America Henry Houston Bonnell of Philadelphia, donated most of his fine collection of Brontë manuscripts, letters and drawings to the newly founded museum.

In the one hundred and fifty years since *Jane Eyre* was first published, millions of visitors have made their way to the British Museum and now the British Library, to view the manuscript of this extraordinarily influential work of literature. Millions more have found their way to the door of Haworth Parsonage, to see where Charlotte lived out her life, whatever biographical version they choose to believe. Many have come to *Jane Eyre* through the relay of movies produced over the years, from Orson Welles' Hollywood version of 1944 to Zeffirelli's restrained interpretation of 1996. But most of all, generation upon generation of new readers throughout the world continue to read, enjoy and read again the powerful and captivating novels of Charlotte Brontë, one of the greatest writers of English literature.

CHARLOTTE BRONTË 1816–1855

Chronology

1816   21 April Charlotte Brontë born at Thornton, Bradford, Yorkshire

1817   26 June Patrick Branwell Brontë born

1818   30 July Emily Jane Brontë born

1820   17 January Anne Brontë born

1820   the Brontës come to Haworth

1821   mother's death, aunt Elizabeth Branwell comes to look after the family

1824-1825 Cowan Bridge School

1825   deaths of sisters Maria (born 1814) and Elizabeth (born 1815)

1826   gift of toy soldiers, The Young Men's Plays begin, developing into the
Brontës' imaginary worlds of Glass Town and Angria

1828   Charlotte Brontë's earliest extant manuscript

1829-1830 art lessons from John Bradley of Keighley

1831-1832 Charlotte goes to Miss Wooler's Roe Head school, Mirfield

1834   Charlotte exhibits drawings at Leeds exhibition

1835-1838 Charlotte is teacher at Miss Wooler's school

1837   correspondence with Robert Southey

1839   May-July Charlotte is governess to the Sidgwicks at Stonegappe,
Lothersdale

1841   March-December Charlotte is governess to the Whites at Upperwood
House, Rawdon

1842   February-November Charlotte with Emily goes to Pennsionat Heger,
Brussels. Both return home on death of aunt Elizabeth Branwell

1843-1844 Charlotte spends second year in Brussels

1846   February, Charlotte sends the manuscript of *Poems* of Currer, Ellis
and Acton Bell to publishers Aylott and Jones

1846   May, *Poems* published

1846 27 June Charlotte completes *The Professor*

1846 August-September Charlotte goes with her father to Manchester where he has an eye operation. Charlotte starts to write *Jane Eyre*

1847 July, Anne's *Agnes Grey* and Emily's *Wuthering Heights* accepted for publication by Thomas Cautley Newby, but not Charlotte's *The Professor*

1847 19 October, *Jane Eyre* published by Smith, Elder and Co., to instant acclaim

1847 December, *Agnes Grey* and *Wuthering Heights* published

1848 7 July, Charlotte and Anne Brontë go to London to prove there is more than one author called Bell

1848 24 September, death of Branwell Brontë, aged 31

1848 19 December, death of Emily Brontë, aged 30

1849 28 May, death of Anne Brontë, aged 29, in Scarborough

1849 26 October, *Shirley* published by Smith, Elder & Co.

1849 29 November, Charlotte visits London as a celebrated author, and meets her hero Thackeray

1850 30 May, Charlotte visits London

1850 3 July, Charlotte visits Edinburgh with George Smith

1850 19 August, Charlotte visits Windermere, where she meets Elizabeth Gaskell

1850 10 December, Charlotte's Preface to the edition of *Agnes Grey* and *Wuthering Heights*, with Biographical Notice about her sisters

1851 28 May, Charlotte visits London

1853 28 January, *Villette* published

1853 22 April, Charlotte stays with Elizabeth Gaskell in Manchester

1853 19 September, Elizabeth Gaskell visits Charlotte at Haworth

1854 29 June, Charlotte marries Arthur Bell Nicholls, honeymoon in Ireland

1855 31 March, death of Charlotte Brontë, aged 38

1857 25 March, Elizabeth Gaskell's *The Life of Charlotte Brontë* published

1857 6 June, *The Professor* published

1861 7 June, death of Patrick Brontë, aged 85

Further Reading

Christine Alexander and Jane Sellars,
The Art of the Brontës,
Cambridge University Press, 1995.

Christine Alexander,
The Early Writings of Charlotte Brontë,
Basil Blackwell, 1983.

Juliet Barker,
The Brontës,
Weidenfeld and Nicolson, 1994.

Elizabeth Gaskell,
The Life of Charlotte Brontë,
Penguin Classics, 1985.

Margaret Smith,
The Letters of Charlotte Brontë,
Volume One, 1829-1847,
Clarendon Press, Oxford, 1995.

T J Wise and J A Symington,
The Brontës: Their Lives, Friendships and Correspondence,
Basil Blackwell, 1932.

Index

The British Library is grateful to the Brontë Society, the Trustees of the National Portrait Gallery, London, the National Maritime Museum, London, Bradford Art Galleries and Museums, York City Art Gallery, the Guildhall Library, Corporation of London, the British Film Institute, Simon Warner, and other named copyright holders, for permission to reproduce illustrations.

Front cover illustrations: Charlotte Brontë by George Richmond, 1850 (courtesy of the Trustees of the National Portrait Gallery); manuscript of the opening of *Jane Eyre*, The British Library Additional MS 43474, f.1; modern view of Haworth Moor (courtesy of Simon Warner)

Back cover illustrations: Charlotte; detail from Branwell Brontë's portrait of his teenage sisters, 1834 (courtesy of the Trustees of the National Portrait Gallery); the earliest known image of Haworth Parsonage, 1850s (courtesy of the Brontë Society)

Half-title page: The Brontës at home, engraving by Joan Hassall, 1953.

Frontispiece: The opening of *Shirley*, Volume 3; *see* page 83.

Contents page: Modern view of Haworth Moor (courtesy of Simon Warner)

Text © 1997 Jane Sellars

Illustrations © 1997 The British Library Board, The Brontë Society, and other named copyright holders

Published in the United States of America by
Oxford University Press, Inc.
198 Madison Avenue
New York, NY 10016
Oxford is a registered trademark of Oxford University Press, Inc.

ISBN 0-19-521439-0

First published 1997 by
The British Library
Great Russell Street
London WC1B 3DG

Designed and typeset by Crayon Design, Stoke Row, Henley-on-Thames
Printed in Italy by Artegrafica